THE PLANETS

Heather Couper is President of the British Astronomical Association. She has had a lifelong interest in astronomy, and graduated with a degree in astronomy and physics from Leicester University. After a period of research into galaxies at the University of Oxford, she became Lecturer at the Greenwich Planetarium, Old Royal Observatory – a post she held for five years. In 1983 she left to concentrate on broadcasting and writing full-time. She has written over a dozen books and appears regularly on radio and TV. *The Planets* is her third television series.

Nigel Henbest is Editor of the Journal of the British Astronomical Association, Astronomy Consultant to *New Scientist* magazine, and Public Relations Consultant to the Royal Greenwich Observatory. After taking a degree in astronomy and physics at the University of Leicester, he obtained a masters degree in radio astronomy at Cambridge, where he studied under the Astronomer Royal, Sir Martin Ryle. He is now an internationally-known writer and broadcaster on astronomy, and his many books include *Mysteries of Space* and *The New Astronomy* (with Michael Marten). He is a researcher on *The Planets* television series.

HEATHER COUPER with NIGEL HENBEST

THE PLANETS

GUILD PUBLISHING LONDON

This edition published 1986 by
Book Club Associates
by arrangement with Pan Books Ltd

This book is linked to the seven-part television series
The Planets produced by **The Moving Picture Company**
for **Channel 4**.

Photoset by Parker Typesetting Service, Leicester
Printed and bound in Great Britain by
Chorley & Pickersgill Ltd, Leeds

ACKNOWLEDGEMENTS

This book could not have been written without the enthusiastic co-operation of dozens of planetary astronomers all around the world. In our travels, we enjoyed long discussions with many of them; we wish there were space to list them here. Our grateful thanks to you all!

There were a number of astronomers who went beyond the call of duty in providing us with information or illustrations. In particular, we would especially like to thank Hal Masursky of the US Geological Survey at Flagstaff for his time, and for the invaluable perspective he gave us. And we'd also like to record our gratitude to Jim Burke (JPL), Jim and Charlene Christy, Henry Fuhrmann (JPL), Gerry Giles (Arecibo), Don Karl (KPNO), Elliot Morris (USGS Flagstaff), Dave Smith (Goddard), Carole Stott (Old Royal Observatory, Greenwich), Don Swann (VLA), Clyde and Patsy Tombaugh, and Peter Waller (NASA–Ames).

Our extensive travels would have been impossible without the backing of Channel 4 and of The Moving Picture Company who produced the television series. It is a pleasure to thank our producer Avie Littler and our director Paul Fisher for putting up with astronomers' jargon for weeks on end.

Our picture researcher, Rose Taylor, took on a mammoth task of ordering 500 illustrations from all over the world, only to see us whittle down the selection to less than 150. And Mandy Caplin once again converted our unreadable longhand scrawl into crisp, error-free type.

Finally, we'd like to record our gratitude to Hilary Davies, our editor at Pan Books. She had faith in the project from the very start.

CONTENTS

To Anita Couper, 1922–1984
With love and gratitude

CHAPTER ONE

NEW WORLDS

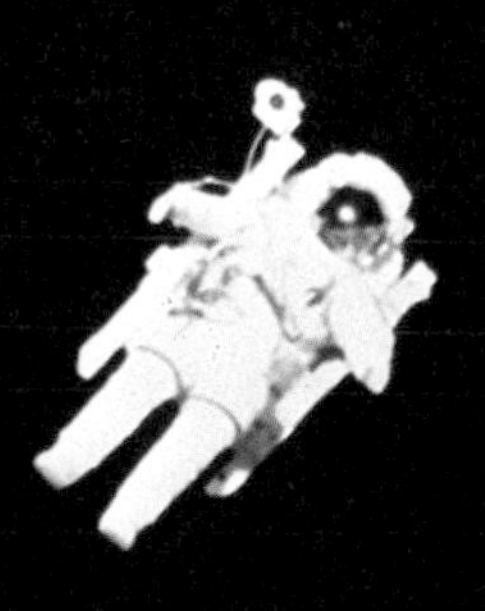

On 11 October 1492, Christopher Columbus and the crew of the *Santa Maria* discovered the New World. It was not the land they had been looking for, but that isn't the point: the fact was that Columbus' bravery and far-sightedness in undertaking such a long uncharted voyage inspired generations to follow his example. A little later, Vasco da Gama reached India; Vasco Nunez de Balboa stowed away and became the first European to set eyes on the Pacific Ocean; Ferdinand Magellan and his crew sailed around the world; and, at about the same time, Hernando Cortes discovered and conquered Mexico. In the intervening 500 years, we have charted, explored and colonized all these 'new worlds' and more. Today, the space shuttle can circle them all in an hour and a half. The new worlds are just continents which we learn about in school geography lessons. Each has its own flavour and character, its own culture and customs, but they are now familiar to us. With almost instant communications and fast travel, we see these once-uncharted territories as parts of a whole – parts of a small world called Earth.

On 12 February 1961, Soviet space scientists launched their first interplanetary space probe towards the planet Venus. They lost contact with it while it was only 4.66 million miles (7.5 million km) out from Earth and the probe missed the planet by 62 000 miles (100 000km), but – as with Columbus' discovery – it didn't matter. The route was now open for others to follow. And today, less than a quarter of a century later, unmanned probes and new, unprecedentedly-sophisticated Earth-based instruments have explored in detail all the planets known throughout history.

These are *real* new worlds. The maps we have of their surfaces are tantalizingly incomplete, luring us on to explore further. And yet, compared with our pre-space era knowledge, what we have learned already is immense. 'Pretty well all we thought we knew was nonsense,' comments space visionary Arthur C. Clarke.

Now that we are discovering the planets as worlds in their own right, instead of just pinpricks of light in the sky, a new unity is beginning to emerge. All these new worlds – individual though they are – make up a tiny corner of our Universe. As we continue to explore, this corner will appear to shrink even smaller, just in the way that our world has. And just as exploring the 'new worlds' of the Renaissance told us a lot about ourselves, so the exploration of the planets will shed light on where we came from – and how we can best protect what we have. Five hundred years from now, our planetary system may seem as accessible to our descendents as the world appears to us today.

Even thirty years ago, such a statement would have seemed crazy. Except to astronomers, the planets were little dots of light which hardly anyone noticed and even fewer cared about. Our new relationship with our fellow worlds has come about so quickly that we have hardly had time to become aware of it. This book is all about these new links, and the directions they will take us in the future.

Looking to the past, though, it's clear that our relationship with the objects in our sky has always been changing – albeit more slowly than now. But exactly when people first tried to make sense of the sky isn't known with any certainty at all. Until 1500 BC, there are no written records. However, the stars and planets were not only watched, but *utilized* well before that. Archaeologists and 'archaeoastronomers' – astronomers who study the prehistory of stargazing on Earth – argue relentlessly over the significance of the mysterious stone circles, avenues and monolithic standing stones found scattered over the granite uplands of northern Europe. Setting aside suggestions that 5000-year-old Stonehenge and its fellows could be 'flying saucer landing pads' or sites where the Earth's 'cosmic energy bubbles up', the consensus is that at least some of these stone artefacts had an astronomical use. Thanks largely to the dedicated groundwork of Professor Alexander Thom and his family, who over many years have surveyed the stones at hundreds of sites along the bleak Scottish coasts and in Brittany, we know that some were aligned with the rising and setting positions of the Sun and Moon. Some astronomers even maintain that Stonehenge was an eclipse computer. At the very least, they claim, the monument could be used as a calendar to monitor the ever-changing rising and setting position of the Sun through the seasons. In this way over many cycles, observers would learn exactly when to expect it to be hidden behind the Moon.

Perhaps we will never know the real function of the stones. In the absence of written records, many archaeologists suspect that we are superimposing our twentieth-century

An Egyptian mummy case of the early 2nd century AD, which shows the goddess Nut surrounded by the signs of the zodiac.

Previous page: *Astronaut Bruce McCandless II floats freely above the Earth, propelled by his hand-controlled backpack. He could circle the entire planet in only 90 minutes.*

concerns and values on the monuments' builders – people who lived fifty centuries ago.

We come on to a surer footing with the Babylonians, whose civilization flourished 3500 years ago. Fortunately, they left written records on baked clay tablets, which revealed that they were keen and accurate observers of the sky. They made rough maps of the star-patterns which appear to be forerunners of those we use today. With crude wooden measuring instruments, they followed the changing position of the Moon and constructed a highly accurate lunar calendar. But one of their main preoccupations was in studying the movements of the 'seven planets', which included the Sun and Moon as well as the five planets we see with the unaided eye today – Mercury, Venus, Mars, Jupiter and Saturn. From flat-topped 'towers of Babel' the astronomers would watch as the planets slowly inched their way, night by night, across the fixed starry background. They noticed that all the wanderers kept to a narrow band in the sky, and divided it up into twelve consecutive zones – each containing one constellation pattern. Thus the twelve signs of the Zodiac were born. To the Babylonian astronomers, the position of a planet along the Zodiac was of vital importance. It was held that these positions had an effect on affairs of state – and in those days of violent political unrest in Mesopotamia, astronomers were greatly in demand for predicting the future!

The Egyptians, who traded with the Babylonians, did not have the same flair for trying to explain *why* the sky changed the way it did, and nor were they very interested in the movement of the planets. But they were meticulous star-watchers. As farmers, the Egyptians were extremely dependent on the flooding of the Nile to provide much-needed irrigation. Their sign that this was imminent came from the stars, coinciding with the first glimpse of the brilliant star Sirius rising in the early autumn dawn sky.

In the East too, the Chinese astronomers kept a close watch on the sky. They were particularly keen to predict the terrifying darkness of total eclipses of the Sun, when they believed that it was being devoured by a dragon. One sad little tale of uncertain veracity concerns the court astrologers Hsi and Ho, who spectacularly failed to warn of the dragon's approach, and were beheaded for their slip-up.

Early astronomy, then – a lot of which involved the changing positions of the Sun, Moon and planets – was a cross between practical timekeeping and speculative astrology. The latter must have grown up as people noticed that certain earthly events, like the Nile floods, always took place when there was the same configuration of the heavens. How magical the astronomers must have seemed to the uninitiated. How clever to predict floods by looking at the stars! Although the heavenly configuration merely reflected the season in the year and not the resultant local weather pattern, the link was established. If the heavens could affect natural events, why not the destiny of whole countries – and even of their individual people?

The first people to separate astronomy from astrology – and thereby put things on a scientific footing – were the Greeks. The still-magnificent remains of their empire, which flourished between 400 BC and AD 100, show how sophisticated and refined their culture was. The same was true of their science and philosophy. Borrowing their basic knowledge of the sky and its constellations from the Babylonians – probably by way of the seafaring Cretan Minoans who traded with them – they then superimposed their own, self-consistent rules. From painstaking observations they constructed a 'model' of any situation, which they reasoned from logical principles. They tested the model with further observations. If these fitted, all was well; but if not the model had to be wrong. The only solution was to come up with a new model, and put that to the test. 2000 years on, this consistent approach – the scientific method – is still used by modern scientists when testing a theory.

If the Greeks had an obsession, it was with symmetry. Symmetry meant perfection; and so the perfect shape was round. This obsession coloured the favourite model of the Universe put forward by the great Greek philosopher Aristotle around 330 BC. The Earth is round, he taught, because you see different stars from different places on Earth. It is fixed in space and perfectly still – otherwise the stars would appear to move past, and there would be violent winds. But things *do* move across the sky, and so they must move around us. And so the Earth is in the centre of the Universe with the Sun, Moon, planets and stars all moving in circles on crystal spheres about it. Finally, all those bodies occupying spheres above that of the Earth – in fact, all the celestial bodies

– are perfect, simply because nobody had ever seen any imperfections. Although Aristotle could never have guessed, his theory of the Universe was to affect the direction of science until nearly 2000 years later.

By the time the Roman Empire had begun to make its presence felt, the Greek civilization was showing cracks. During a lull in the situation during the first century AD, Greek scholars seized the chance to collate their findings for posterity in the great library and museum in Alexandria, Egypt. One of those scholars was Claudius Ptolemaeus (Ptolemy) who collected together the essence of the Greek astronomical findings in his superb encyclopedia the *Syntaxis* – better known by its later Arab name of the *Almagest*.

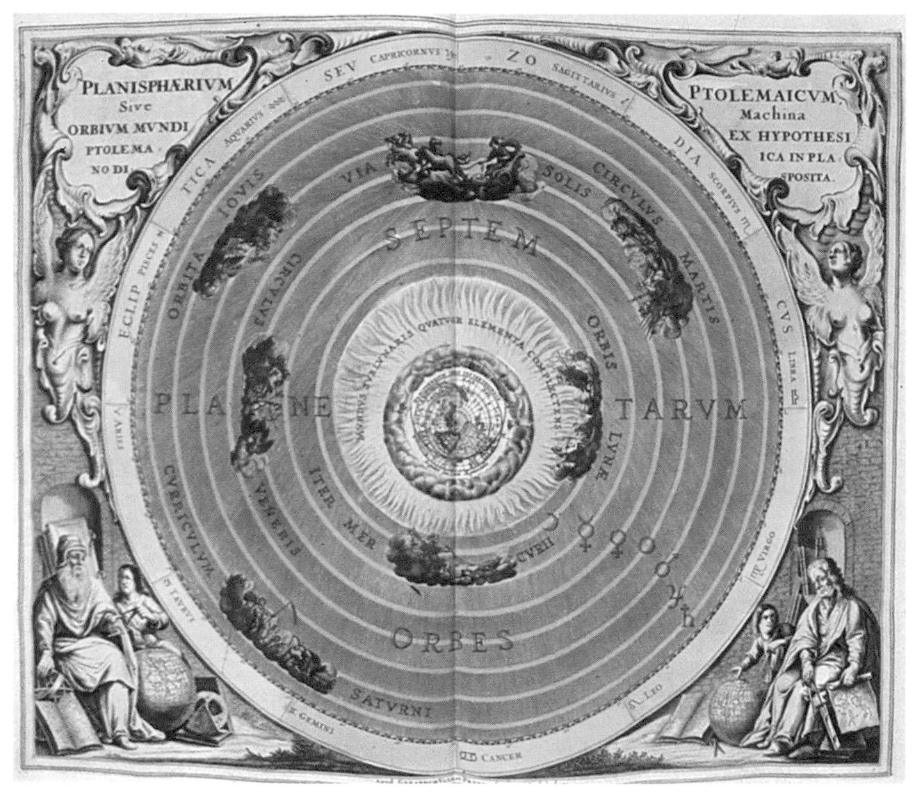

The Ptolemaic theory, which endured for 1500 years, held that the Earth lay at the centre of the Universe. Here it is shown surrounded by spheres of water, air and fire, and by the planets and the stars.

Ptolemy noticed a small fly in the ointment of Aristotle's elegant universe. If the planets circled the Earth on crystal spheres, why was it that they sometimes moved so oddly in the sky? Why did Mars, Jupiter and Saturn occasionally stop their normal westwards motion and then loop *backwards* – towards the east – before resuming their normal course? How was this consistent with the perfect spherical universe?

Ptolemy came up with an ingenious modification. Anxious not to dispose of the perfect circles, he proposed instead that each planet does indeed move in a wide circle about the Earth. But to explain the loop-the-loop behaviour in the sky, he added, a planet must *also* pursue small-scale circular loops (*epicycles*) around its average path. It was a very neat compromise – and Ptolemy was confident enough of his theory to predict the positions of the planets in the sky for many years ahead.

The years ahead were not good ones for the Western world as it plunged towards the Dark Ages. But some of the ancient teachings survived in Arab lands where they were put to good use, and by the ninth century AD, Baghdad had become a major astronomical centre. Even today, we find this reflected in star names – Algol (the demon), Deneb (tail) and Betelgeuse (allegedly 'the armpit of the sacred one'!). With the coming of the Crusades, the Greek texts started to trickle back into Renaissance Europe.

They couldn't have come at a better time. While Renaissance scholars were making far-reaching strides in art, literature and music, they had no tradition in science to build on. With the Greek teachings it suddenly appeared, ready-packaged. But unfortunately, things went wrong. Because the new learning had developed among scholars in the monasteries, the Church in fifteenth- and sixteenth-century Europe was all-powerful. Determined to maintain its rich, pre-eminent position, the Church twisted Aristotle's perfect vision of the Universe into a dogma calculated to uphold the Church's inalienable right to its supremacy. Any challenge to the dogma was heresy, punishable by death.

Ironically, the rescue from this straitjacket came from within the Church itself. An elderly Polish monk, Nicolas Copernicus, had made astronomy his hobby. Over many years, he had become puzzled with the movements of the planets; for, try as he might, he couldn't reconcile the positions in which he observed them with the predictions in Ptolemy's *Almagest*. In order to match theory and reality, he would have to have added many extra cumbersome epicycles to each planet's path. How much simpler, he

The Ptolemaic system was eventually displaced by Copernicus' idea that the Sun lay at the centre of the Universe. In this eighteenth-century Dutch engraving, Copernicus himself is the figure at lower right.

thought, to replace the Earth with the Sun at the centre of the Universe. If the Earth moved around the Sun, it would clear up all kinds of discrepancies.

It's not entirely clear whether Copernicus really believed that the Earth should be demoted to the status of a mere planet, or whether his switch was a mathematical dodge. But he may well have known how much his ideas would anger the Church authorities, for his suggestion (in the form of a book, *De Revolutionibus Orbium Coelestium* – 'Concerning the Revolutions of the Heavenly Bodies') was not made public until he died in 1543.

Predictably, the Church was annoyed – although not greatly, for the idea was that of just one man. But news got around. And the effect it had was to stimulate scientific inquiry at last. People began to make measurements in the sky again, 'just to check'.

Tycho Brahe's meticulous measurements were used by Johann Kepler to prove that the planets orbit the Sun in elliptical paths. Here he is seen in his observatory at Uraniborg, with the enormous mural quadrant he used for his measurements.

Many of the new observers were firmly on the side of the Church. Tycho Brahe, an eccentric Danish nobleman who in 1576 built the world's most sophisticated observatory on the island of Hven, hoped very much that his meticulous observations would uphold the supremacy of the Earth. Working with enormous measuring arcs and circles – for this was just before the invention of the telescope – Tycho measured the positions of the wandering planets to a degree of accuracy never before achieved. But when his assistant, the brilliant mathematician Johannes Kepler, came to analyse them, the Church received a double blow. Not only did they support Copernicus' theory; they also revealed that the planets didn't even travel in perfect circles! Instead, they appeared to move around the Sun in egg-shaped paths, or ellipses. In his three 'laws of planetary motion' Kepler demonstrated that a planet would move faster when it was closer to the Sun in its elliptical orbit, and slower when further away. Also, the planets closer to the Sun moved faster than the distant ones, giving a natural explanation as to why Mars, Jupiter and Saturn occasionally looped backwards – it was simply that the Earth, in its faster inner orbit, was overtaking them.

Galileo was a scientist and mathematician of high ability. Quite apart from his discoveries with the telescope, he contributed greatly to the fields of ballistics and dynamics.

Into the controversy came the telescope. No one is sure who invented it: but it seems that by the beginning of the seventeenth century, the Dutch (and possibly the English) were familiar with the fact that a combination of two particular kinds of lenses would make distant objects appear closer. It was Galileo Galilei, professor of mathematics at Padua, Italy, who first turned such an 'optik tube' towards the skies and interpreted what he saw.

Later chapters will reveal more of what Galileo saw. But his discoveries were enough to convince him – very firmly indeed – that Copernicus and Kepler were right. Through a tiny telescope no bigger than a modern toy, Galileo observed spots on the Sun and mountains on the Moon – proof that they were not 'perfect'. He saw that Venus had a cycle of phases like those of the Moon, changing from crescent to full and back again – and it did so in a way which showed that it circled the Sun, and not Earth. Jupiter was surrounded by four little points of light which moved around it from night to night – evidence that not *everything* was in orbit about *us*.

Tempestuous, red-haired Galileo clashed violently with the Church. The authorities felt so threatened by his assaults on their dogma that, in the end, they put him under virtual house arrest for the last nine years of his life. Half-blind, yet with his spirit quite unbroken by the trials and inquisition he had gone through, Galileo ended his days researching into the problems of moving bodies and mechanics.

News of Galileo's discoveries spread rapidly to northern Europe, where the hold of the Church was far weaker. In 1667, just twenty-five years after Galileo's death, the French founded a national observatory in Paris, and England's Royal Greenwich Observatory followed in 1675. Even before this, a group of scientifically-motivated gentlemen had founded 'The Royal Society of London for Improving Natural Knowledge', which still survives today as the Royal Society. The aims of the Society, founded in 1660, were specifically to verify theories of every kind by *experimental* evidence – rather than testing ideas by turning to ancient authority.

One of the first illustrious presidents of the Royal Society was Isaac Newton, who was born in the year of Galileo's death. It was his Theory of Gravity which finally supplied the missing pieces of the jigsaw. Newton showed that every object in the Universe has a gravitational pull which attracts other objects to it. The more massive the object – the more material it contains – the greater is its gravitational pull. And the further you go from an object, the weaker that gravitational pull becomes.

Here, at least, was the reason why the planets go around the Sun – and at different speeds, as Kepler had found. The Sun is so massive that it keeps all the planets securely in tow. If they didn't circle the Sun, they would be pulled straight in. But those further out can afford to go at a more leisurely pace, because the pull they feel is weaker.

And so it took only a hundred years to change the long-held certainty that the Earth sat at the centre of the Universe to a realization that it was somewhere out on the sidelines. Over the three centuries which followed, with the new no-holds-barred attitude to scientific enquiry, Earth appeared less significant with every new finding. Since Newton's day, three new planets have been found, beyond the ones known to the Babylonians and Egyptians. In the nineteenth century, we began to measure the distances to the stars, expanding our frontiers a million times over. This century, we have been able to probe galaxies – 'cities' of stars – a million times further away again. And our frontiers are still widening as new techniques, new instruments, come into play.

Galileo's telescopes were no larger than modern toys. The discoveries he made with them, however, revolutionized our thinking about the Universe – largely because Galileo was a very perceptive interpreter of his observations.

We now know that the Sun and its nine orbiting planets comprise a group in space, called the Solar System. In order from the Sun, the planets are Mercury, Venus, Earth, Mars, Jupiter, Saturn, Uranus, Neptune and Pluto – usually. From 1979 until 1999, the order is actually Pluto and *then* Neptune, because Pluto has a peculiar orbit which carries it closer to the Sun than its neighbour for a brief period.

Small though the Solar System is on the cosmic scale of things, it is vast by human standards. The distance from the Sun to the Earth is ninety-three million miles (150 million km). It would take over twenty-one *years* to reach the Sun by transatlantic jet. The same jet would require almost 850 years to make the journey to Pluto. Like most of the Universe, the Solar System is mainly empty space. Sprinkled throughout – and particularly between the orbits of Mars and Jupiter – there is debris left over from its birth, the asteroids and meteoroids. Much further out, although no one knows exactly where, there may lie a huge cloud of snowballs which totally surrounds the Solar System. This cloud is the source of the comets which have terrorized superstitious people through the centuries.

The planets themselves have one thing in common: they are all dark, and shine in our skies only because they reflect sunlight (like a high-flying aircraft seen at dusk). Otherwise, each has its own emphatic personality. But broadly speaking, the four planets near the Sun – Mercury, Venus, Earth and Mars – share a family resemblance in being small, solid and rocky. Likewise, Jupiter, Saturn, Uranus and Neptune are not dissimilar. Each is a huge gas-ball with no 'surface' at all, surrounded by retinues of encircling moons, and rings. Pluto, the smallest and most recently-discovered planet, fits into neither of these camps.

Dominating the entire Solar System is the Sun. Nearly a thousand times bigger than all the planets put together, the Sun is very different from the rest. It is not a planet at all, but a star – our local star. Instead of being cold and dark, it is hot and dazzlingly bright. The reason for the difference lies in the Sun's enormous mass, 330 000 times that of the Earth. Under the weight of the outer layers, the Sun's central regions are so hot and compressed that nuclear fusion reactions go on at its heart. In effect, the Sun's core is a gigantic hydrogen bomb – but the weight of those outer layers will prevent the reaction from ever running out of control.

The Sun, along with all the other stars, held the secret of energy production by nuclear fusion long before it was learned by us. That energy, created deep down in its core where the temperature is fourteen million °C, bubbles to the Sun's surface as light and heat. Even the Sun's 'surface' – the topmost layer of opaque material – has a temperature of 5500°C. The whole of the Sun is so hot that nothing solid could exist there, and its entire vast globe is made of gas.

The Sun's effects on its surrounding worlds go further than gravity. Its gravitational pull is by far the most important, keeping the planets in their place, but its other influences are considerable, too. There is the Sun's energy for a start – the heat and light upon which the planets depend for developing their own sources of energy. Then

This imaginative mural, more than 12 feet across, shows the sizes of the Sun and the planets to the correct scale. Only part of the Sun can be represented here, while the Earth-Moon system is difficult to see at all.

Astronauts aboard the Skylab space station photographed this 'artificial eclipse of the Sun' by blotting out its brilliant disc. Here, the Sun's thin outer atmosphere – colour-coded for its brightness – can be seen stretching millions of miles into space.

there is the fact that all planets actually circle the Sun within the outer layers of its atmosphere. Look at the Sun during a total eclipse – when our Moon completely hides its brilliant disc – and you will see an encircling pearly halo fading into the twilight sky. This, the Sun's corona, is the origin of the solar wind, a stream of charged atomic particles which blows from the Sun throughout the Solar System. When the Sun's 'weather' is stormy, and its smooth surface is speckled with dark, cooler sunspots, the wind can rise to gale-force – and the planets feel the effects. But despite periodic bouts of stormy weather, which happen roughly every 11 years, our Sun is a dependable star. There are others we literally couldn't live with.

In fact, our Sun is remarkable in its ordinariness. It comes middle-of-the-league in everything – size, mass, temperature, brightness. This we know from being able to compare it with stars which lie close by.

But even the nearest neighbouring star, Proxima Centauri, is twenty-five million million miles (40 million million km) away. A transatlantic jet starting off on this journey would not reach its destination until six million years later.

Yet this is only the beginning. On a brilliantly clear night, you can pick out 3000 stars in the sky. Binoculars reveal thousands more, and telescopes can photograph stars running into their millions. All the stars we see – and countless more we can't see – go to make up our home 'star-city', the Milky Way Galaxy. When you see the misty band of the Milky Way flowing through the stars on a clear night, you are looking into the thickness of our lens-shaped system. On looking to either side you see out into the depths of empty space. The Sun's position in the Galaxy is not at all privileged; we live about two-thirds of the way out from the centre.

Our Galaxy contains over 100 thousand million stars. It is so vast that our jet would take over 100 thousand million years to cross it. And beyond our Galaxy there are many, many others, each with their own complement of stars. Galaxies stretch as far as our present instruments can probe. There may be more than 100 000 million million million stars in all the galaxies which make up our Universe.

So, in relation to the whole known Universe, the Sun and its family of planets seem utterly insignificant. Stars like the Sun are more than ten-a-penny – and so we would expect planets like our own to be equally commonplace. We and our fellow worlds make up an average corner of the cosmos; and by getting to know our neighbourhood well, we will arrive at a better understanding about how the Universe as a whole works.

We will also learn more about our own world. Seeing it against its hostile neighbour planets makes us realize how fortunate we are. But it is up to us to decide whether we want to run the risk of Earth suffering a parched, arid future, like Venus, or a permanent ice age, as on Mars. It all depends on how well we treat our planet in the forthcoming century. Both these possible fates have only become obvious since we began to get to know our fellow planets as worlds in their own right.

But this is not a book about doom and despondency. It is a celebration of nine unique worlds in space, each with its own special character. It is a geography lesson about the different 'continents' which make up our much larger 'world' – the world of the planets. And it's a book which couldn't have been written even a few years ago. At the moment, as space probes prepare to investigate our neighbour worlds in more and more detail, we are undergoing a revolution in our knowledge which may turn out to be as far-reaching as the one Copernicus started. Already, there is one thing which is certain. We have learnt more about the planets in the last twenty years than throughout the whole previous history of mankind.

CHAPTER TWO

MERCURY

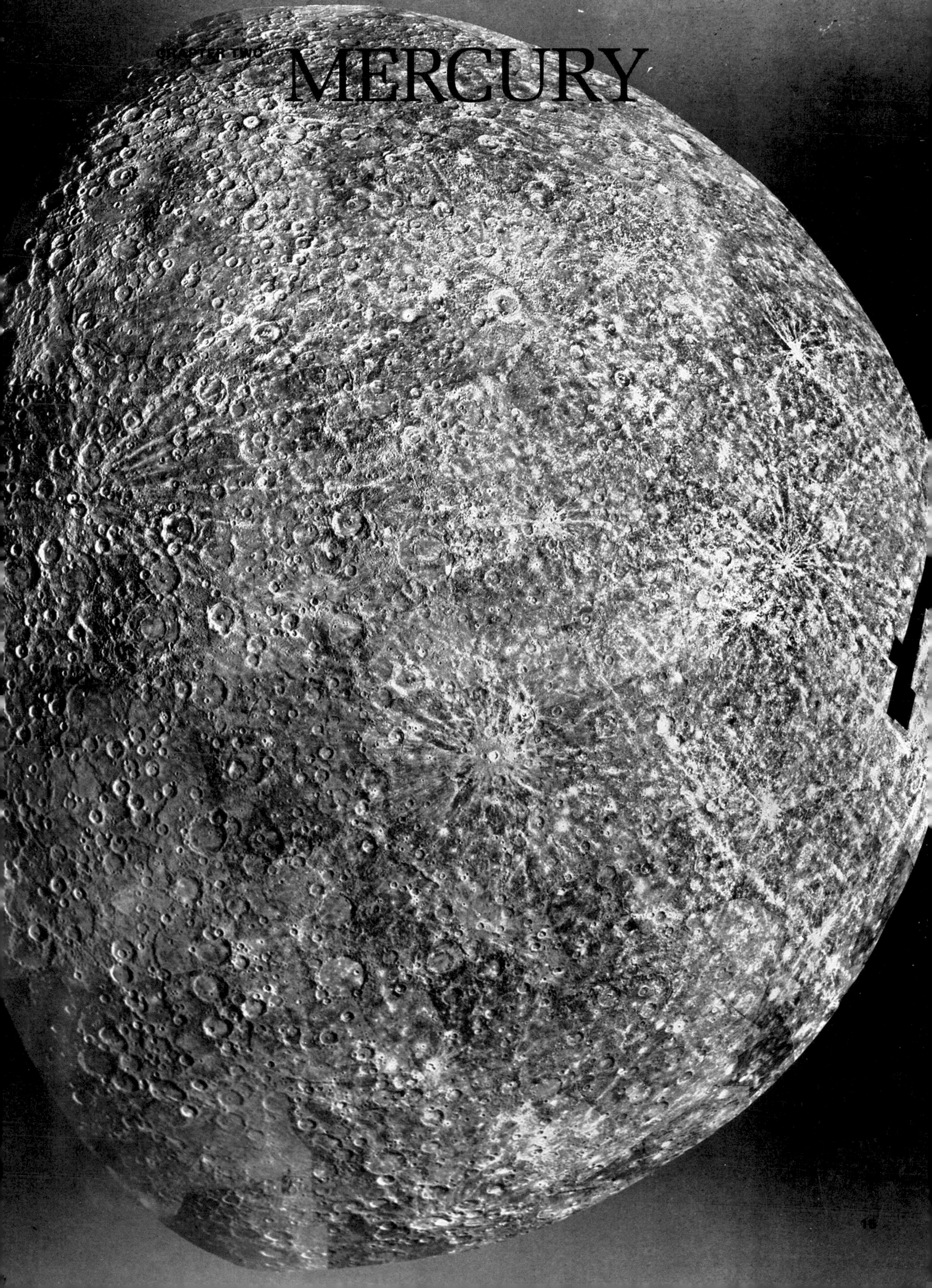

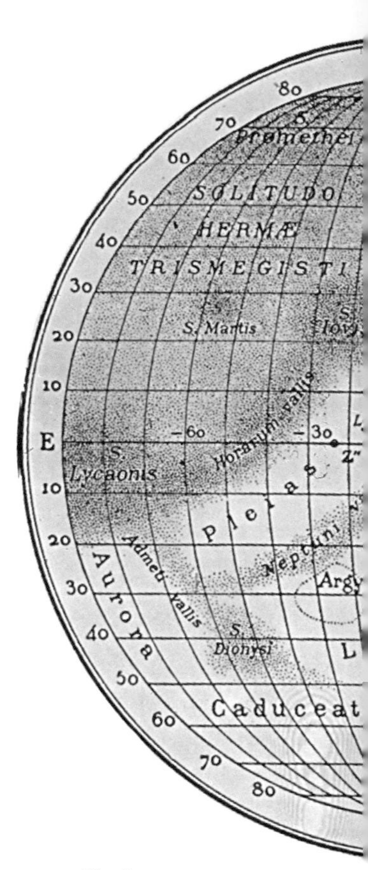

Previous page: *The barren, crater-splattered surface of Mercury has remained unchanged since the first hectic years of the Solar System. This photograph was taken by the only spaceprobe to go there, Mariner 10.*

This wall-painting from 1st-century Pompeii shows Mercury as the fleet-footed messenger of the gods. His 'caduceus' (herald's staff) was entwined with white ribbons, which were sometimes interpreted as serpents.

'Mercury is a shy visitor, for he follows the Sun more closely than his neighbour Venus and is, besides, much smaller; he is, therefore, not nearly so familiar an object in the sunset sky, and even when observed is often not recognized.' Lucy Taylor's delightful introduction to the innermost planet, from her 1895 book *Astronomers and their Observations*, is perhaps even truer today than it was then. With the spread of streetlighting, Mercury – always very low in the sky – is often drowned out completely, and most people have never seen it.

Unlike the planets which orbit the Sun beyond the Earth, Mercury – along with its neighbour, Venus – can never be seen high in the sky at the dead of night. Because they circle the Sun more closely than we do, Mercury and Venus never stray far from it in the sky. This means that we only get the chance to see these planets either immediately after sunset, as they follow the Sun down, or preceding the Sun before sunrise. To people living in the country, Mercury and Venus have always been 'the morning stars' or 'the evening stars' – the 'stars' which appear before all the others, before the sky is properly dark.

Mercury, however, can never rival Venus. On average, it's only a fiftieth as bright; and while Venus sometimes seems like a Chinese lantern hanging in the sky, Mercury, at

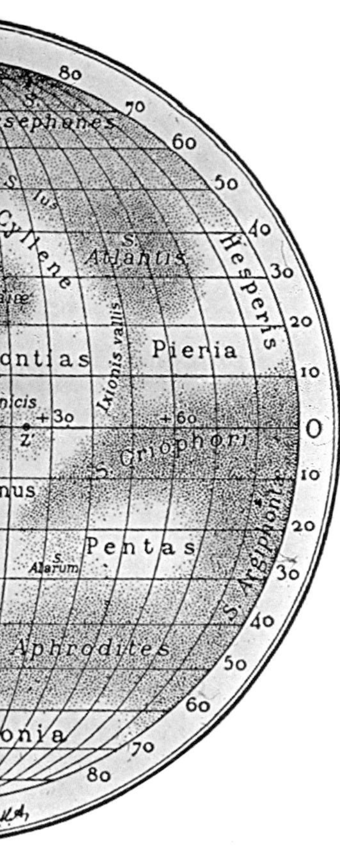

The world's greatest telescopic observer of Mercury, Eugene Antoniadi, saw dusky markings on the planet, to which he gave romantic names. Close-up views from Mariner 10, however, revealed that none of these markings was real.

best, looks like a bright star. But a planet never looks *exactly* like a star. Stars appear to twinkle as hot and cold currents in our atmosphere continually shift in front of them and distort their images. Close to the horizon, where the layers of air are thickest, the twinkling effect is greatest. A really low-lying star appears to turn on and off, and if it's a bright star, it may flash all the colours of the rainbow, too.

Planets, on the other hand, shine with a much more steady light. That's because they're close enough to appear as tiny discs. Stars are so remote that they are seen only as points of light, and so Earth's churning atmosphere can make their light appear to come and go. The disc of a planet, small though it is, is never as obviously affected. Although Mercury does twinkle when it's close to the horizon, its light is noticeably steadier than that of a star at the same altitude.

Distinguishing Mercury from a star is one thing; actually locating it in the sky is another altogether. Aptly named after the fleet-footed messenger of the gods, Mercury is the fastest-moving planet of all. When it does pull out far enough to either side of the Sun to become visible – either just after sunset or just before sunrise – it's only a matter of days before it plunges back again into the Sun's glare. It can only be glimpsed on a handful of occasions in the year. The best times to hunt for it are after sunset in the early spring, and before sunrise in the autumn – but you will need to check with a starchart and a handbook giving the planets' positions to be absolutely sure.

Small wonder, then, that several astronomers have lived out their lives without ever seeing Mercury. Among them was Copernicus, who, it is said, never saw the planet because he was plagued by mists rising from the River Vistula which flowed past his home. Until the spaceprobe Mariner 10 flew by it, almost nothing was known about Mercury except its size – roughly a third that of the Earth – and the length of its year. As the closest planet to the Sun, Mercury takes only eighty-eight days to complete an orbit, travelling at more than 110 000mph (170 000kph). The reason for this breakneck speed is the Sun's immense gravitational pull. At only a third the Earth's distance from the Sun, Mercury – like a conker whirling around on a short string – has to travel very fast indeed to avoid being pulled right in.

This fast-moving, elusive little world, only fractionally bigger than our Moon, presented a challenge to past generations of astronomers. The first telescopes were too small to reveal any of its features at all. But as larger telescopes, capable of collecting more light, came on the scene, finer details began to emerge. Late in the seventeenth century, the Polish astronomer Johannes Hevelius noticed that Mercury showed a phase like that of the Moon. At its closest, the planet showed a thin crescent. Then, as it drew away in its orbit round the Sun, the amount of the illuminated portion grew until Mercury was almost 'full'. Unfortunately, this meant that very little of Mercury's disc was illuminated by sunlight when the planet was at its closest to the Earth.

The giant 1000-ft diameter radio telescope at Arecibo, Puerto Rico, first measured Mercury's rotation period in 1965. Until then, it had been thought that Mercury kept one side turned permanently towards the Sun.

But Giovanni Schiaparelli, better known as the man who first sketched the Martian 'canals', reckoned that he could *just* pick out streaky markings on the planet's surface. From a series of observations beginning in 1881, he concluded that Mercury's 'day' was the same as its 'year' – so that it always keeps the same face turned towards the Sun, just as the Moon keeps the same face towards the Earth. This wasn't too surprising: with Mercury being so close to the Sun, astronomers had expected its spin to have been 'braked' by the Sun's powerful gravity.

The world's most respected observer of Mercury, Eugene Antoniadi, agreed with Schiaparelli. For five years, between 1924 and 1929, he made detailed maps of the planet through the giant 33 inch (83cm) diameter lens of the Meudon telescope, situated just outside Paris. His maps show dusky, greyish markings to which he gave romantic names from Egyptian and Greek mythology. In Mercury's southern hemisphere, for instance, we find Chaos, Hesperis, and Solitudo Hermae Trismegisti – 'Wilderness of Hermes the Thrice Greatest'. Sadly they didn't survive close-up spaceprobe scrutiny.

Even before the Mariner 10 probe reached Mercury, there'd been a little more progress made in understanding the quicksilver world. In 1965, astronomers decided to check whether Mercury spins or not by a novel method – a kind of interplanetary sonar. Using the word's largest radio telescope – the 1000 foot (305 metre) dish at Arecibo, Puerto Rico – astronomers aimed powerful radio pulses towards the planet and carefully analysed the echoes reflected back from its surface. And as in conventional sonar, where echoes received back on ship tell of the sea floor's structure, so the radar echoes from Mercury's surface revealed the planet's contours. They also showed that Mercury *is* rotating.

Mercury is now known to spin once every 58.65 Earth days. This is exactly two-thirds of Mercury's 'year' of 87.97 days. As a result, one complete Mercury 'day', from sunrise to sunrise lasts 176 Earth days – two orbits of Mercury around the Sun. Having a 'day' longer than a 'year' is confusing enough, but the true situation is even more complicated. Pluto excepted, Mercury has the most elliptical orbit about the Sun of all the planets. At the most distant point in its orbit – the aphelion – Mercury is half as far away again as when it is at its closest (perihelion). Depending on its distance from the Sun, which dictates how strong a pull of gravity it feels, Mercury's speed in orbit is always changing. When it's at perihelion, it travels fastest – which leads to a very strange effect. At this point. Mercury's spin on its axis is actually slower than its speed around the Sun. As seen from the planet's surface, the Sun would appear to stop its slow westward drift in the sky; for just a few hours, it would actually move across the sky from west to east. Then, once Mercury had moved on to a 'slower' part of its orbit, the Sun would resume its normal east-to-west path above the planet's arid, heat-baked surface.

Mariner 10 is the only spaceprobe ever to have visited this scorched little body. Launched by the United States towards Venus in November 1973, it was the first probe to pass two planets, with Mercury as the main target. Passing Venus only three months after launch, Mariner 10 was dramatically braked and deflected sunwards by the planet's gravity. After this meticulously calculated game of interplanetary billiards, the probe continued on course for Mercury.

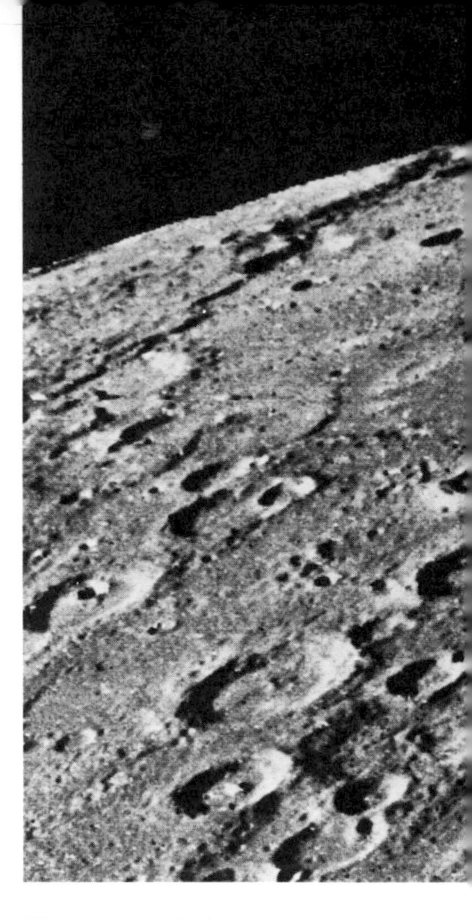

This Mariner 10 photograph of the north pole of Mercury shows a number of surface 'wrinkles' stretching for hundreds of miles. They are probably a result of Mercury shrinking, so that its 'skin' is now too big for it.

The mission was carefully planned so that Mariner 10 would get more than just one peek at the little planet. After passing Mercury's night side at a mere 430 miles (690km), the probe went into orbit around the Sun – in effect becoming a tiny planet in its own right – to ensure that it would pass Mercury (or 'encounter' it in space jargon) at least once more. In the event, it did so twice – in September 1974 and again in March 1975.

Both times, Mariner 10's cameras and sensors transmitted back images of a world almost identical to our Moon. Not only are Mercury and the Moon similar in size, but they are both absolutely covered in craters of all sizes. The two are so alike that, given a close up of a small region of surface, it is difficult to tell them apart. But there are subtle differences. Like the Moon, Mercury's craters are the result of an intense period of bombardment by meteoroids (leftover chunks of debris) which took place early in the history of the Solar System. Mercury, however, although similar in size to the Moon, is more than four times as massive, with a stronger gravitational pull. This means that the ejecta blankets, the material thrown out when the craters were formed, cover a much smaller area on Mercury, because the ejected debris is not able to travel as far. But the craters themselves, although shallower than their lunar counterparts, span a similar size range. These are a few truly colossal scars, some more than 125 miles (200km) across.

Seen from Mariner 10 at a distance of 124,000 miles (200,000km), Mercury looks almost identical to our Moon. It is covered with impact craters of all sizes, the largest being 125 miles (200km) in diameter.

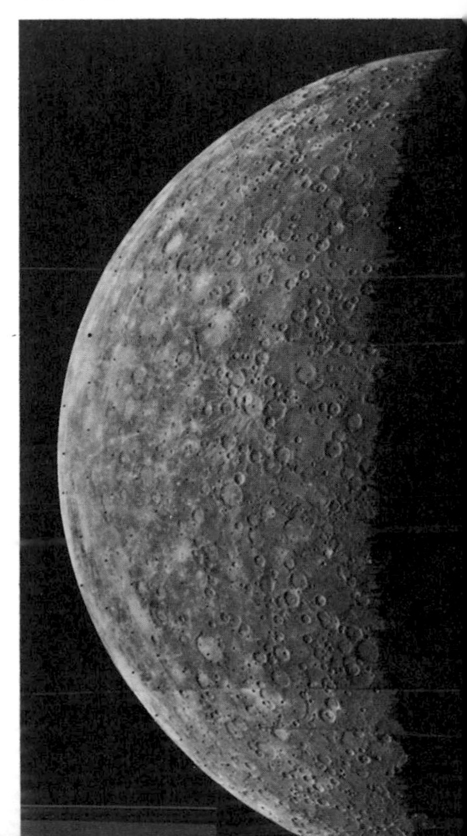

The biggest scar of all is the 800 mile (1300km) diameter Caloris Basin, with its concentric rings of mountains. There are similar wounds on the Moon (the Mare Orientale) and on Saturn's moons Mimas and Tethys. All of them are the result of direct hits by chunks of debris which must have been hundreds of miles across. The chunk which blasted the Caloris Basin made waves over the whole of Mercury, and it's even thought that the tremors from the impact threw up little ranges of hills on the opposite side of the planet. Today, the area is quiet and flooded with cracked, solidified lava. Ever since the planets finally 'mopped up' almost all the leftover debris – nearly 4000 million years ago – Mercury's surface has been left in peace.

But there's another twist to the Caloris story. Most of the features discovered by Mariner 10 on Mercury have been named after famous men. Thanks to a special committee of the International Astronomical Union, we find Bach, Beethoven and Wagner; and Shakespeare, Tolstoy and Homer. Appropriately, one of the most prominent craters has been named after Gerard P. Kuiper – the father of modern planetary astronomy. Caloris, however, has no cultural associations: it got its name simply because it is the hottest spot on the planet. The Basin is situated on the part of

Mercury where the Sun shuttles back and forth overhead when the planet is at perihelion. As Mercury is completely airless – it's too hot and too small to hang on to an atmosphere – the temperature here at noon can climb to 415°C. An astronaut brewing a cup of tea in the Caloris Basin would have little success: his tin kettle would simply melt! But Mercury's lack of air also makes it a planet of great extremes. As there's no atmosphere to keep the heat in, it cools very rapidly when night falls. A Mercurian night is a bitter affair, with temperatures plunging to −170°C.

The temperature extremes and the battered surface of Mercury seem to make it a world apart from our equable Earth. And yet deep down, the two planets may be more alike than first meets the eye. We already know that Mercury is a lot more massive than the Moon, and so – since they're nearly the same size – they must have different kinds of interiors. In fact, in its overall density, Mercury is much more similar to the Earth than the Moon. Earth's surface layers of rock surround a much denser core of liquid iron: does Mercury, then, have an iron core too?

One of the big surprises of the Mariner 10 mission was the discovery that Mercury has a very weak magnetic field. The Earth has a magnetic field, too – but in our case, it's not that difficult to explain. Earth's field is thought to have its origin deep-down in its fast-spinning iron core, which works rather like a dynamo generating electricity. But where does slow-spinning Mercury's field come from? The only explanation is that it somehow arises from a similar iron core inside the planet. But in Mercury's case, the core is proportionally very much bigger: possibly as much as 80 per cent of the size of the whole planet.

If Mercury's core *is* this big, then it is probably still cooling down since the hot, hectic days of the planet's formation. As time goes by, the liquid metal will slowly solidify. And since solid metal takes up less room than liquid metal, Mercury's core has shrunk.

Like the skin around an old, shrunken apple, the thin crust of Mercury has crumpled up to keep pace. Mariner 10 saw this in the shape of features which have no counterpart on the Moon: long, winding ridges hundreds of miles long, rising to 9800 feet (3000 metres) in places – as high as the Pyrenees. These ridges have no regard for the underlying terrain. They run over craters and plains alike, pinching the already-existing surface wherever they touch. These 'wrinkle ridges' may be the outward signs of a shrinking planet. If this is the case, then Mercury need only have contracted a couple of miles since its birth to produce the wrinkle ridges we see today. The innermost planet is indeed a wizened little world.

But *is* Mercury the innermost planet? Might there be a planet even closer to the Sun? A century ago, a number of astronomers were quite certain that there was; and they named it – most appropriately – after Vulcan, the Roman fire-god.

Rings of concentric mountain ridges thrown up by Mercury's biggest meteorite impact surround the 800-mile (1300km)-wide Caloris Basin. The basin's name stems from the fact that it is the hottest place on the planet.

One compelling reason for believing in Vulcan was Mercury's strange orbit about the Sun. As we already know, it is very elliptical, or egg-shaped. However, the long axis of the 'egg' doesn't stay pointing in the same direction, but slowly swings around instead. This means that, over many years, Mercury's real route through space makes up a complicated rosette pattern around the Sun. This kind of behaviour goes by the cumbersome title of 'perihelion precession'.

Urbain Leverrier, the Director of the Paris Observatory, was one of the staunchest believers in the intra-Mercurial planet Vulcan. Here he is seen depicted on the obverse of a medal struck in 1884, seven years after his death.

To nineteenth-century astronomers like Urbain Leverrier, Director of the Paris Observatory, Vulcan was to blame. Leverrier had just successfully predicted the position of the newly discovered planet Neptune from the pull of its gravity on Uranus. Now, here was Mercury behaving in a peculiar way. If there *was* a planet closer to the Sun, then surely *its* gravity would pull Mercury out of position in just the same way?

The difficulty, of course, would lie in finding it. If Mercury was elusive, then Vulcan – closer in to the Sun – would prove more elusive still. But there were two possible ways to locate it. One would be during an eclipse of the Sun, when the Moon blots out the Sun's bright disc. Any intra-Mercurial planet would show up as a tiny 'star' very close to the eclipsed Sun. The other would be if someone were to observe a transit of Vulcan, with the planet seen in silhouette against the Sun's bright disc. Only planets closer to the Sun than Earth can undergo transits, and those of Venus in particular were once the only way of establishing the size of our Solar System. Transits of Mercury, in fact, are relatively common, and there will be three more this century – the first on 13 November 1986.

When, in 1859, Leverrier heard that an unknown French amateur astronomer had reported a small dark spot crossing the face of the Sun, he became very excited. It couldn't be a transit of either Mercury or Venus: it simply had to be the missing Vulcan. Leverrier made up his mind to visit the amateur and see if he was speaking the truth. The meeting must have been a memorable one – impatient, irascible Leverrier confronting the eccentric country doctor Edmond Lescarbault, who, faced with a shortage of paper, recorded his astronomical observations on a plank of wood and planed them off when he had no further use for them. Pretending to be sceptical, Leverrier bullied poor Lescarbault about 'the grave offence of keeping your observations secret for nine months' before he was finally satisfied with the man's claims.

Delighted, Leverrier was now convinced of Vulcan's existence. From Lescarbault's observations, he calculated that it must lie 13 082 000 miles (21 053 000km) from the Sun – only about a third of Mercury's distance. But other astronomers were steadily becoming more sceptical. Very careful observations during the eclipse of 29 July 1878 (a year after Leverrier's death) failed to turn up any sign of Vulcan. Some astronomers later checked their records of the Sun on the day that Lescarbault had allegedly discovered the new world, and found nothing.

It was Lescarbault himself who finally laid Vulcan to rest. In 1891, he announced his discovery of a 'new star' in the constellation of Leo to the French Academy of Sciences. The 'star' turned out to be the planet Saturn. There can be no remaining doubt that poor Lescarbault was just an observer of very little competence – and what he saw on the Sun as 'Vulcan' may well have been simply a defect in his telescope.

Mercury's puzzling 'rosette' orbit remained unexplained long after people stopped believing in Vulcan. Albert Einstein's General Theory of Relativity finally supplied the answer in 1915, when he proved that Mercury's peculiar motion was due to the powerful gravitational field of the Sun.

But what about the strange behaviour of Mercury's orbit, if there's no Vulcan to pull it out of position? To solve the mystery, we must move forward to the early years of this century – to Albert Einstein. In 1915, Einstein had just completed his General Theory of Relativity. This is basically a theory of gravity which describes what happens to objects in gravitational fields which are very strong. In weak gravity – which suits the situations we come up against in everyday life – it's fine to use the familiar theory of gravity worked out by Isaac Newton. But under strong gravity, such as that felt by Mercury close to the Sun, bodies behave according to Einstein's law. And according to Einstein, Mercury *shouldn't* follow the same oval again and again. The oval orbit should gradually swing forward – just as we observe it doing. In fact, Mercury follows Einstein's predictions so well that there *can't* be a Vulcan – unless it is very small.

And so Mercury stands alone at the scorched inner frontier of the Solar System. And things look likely to stay that way. Mariner 10 has shown us what we want to see, and we are in no hurry to return.

CHAPTER THREE

VENUS

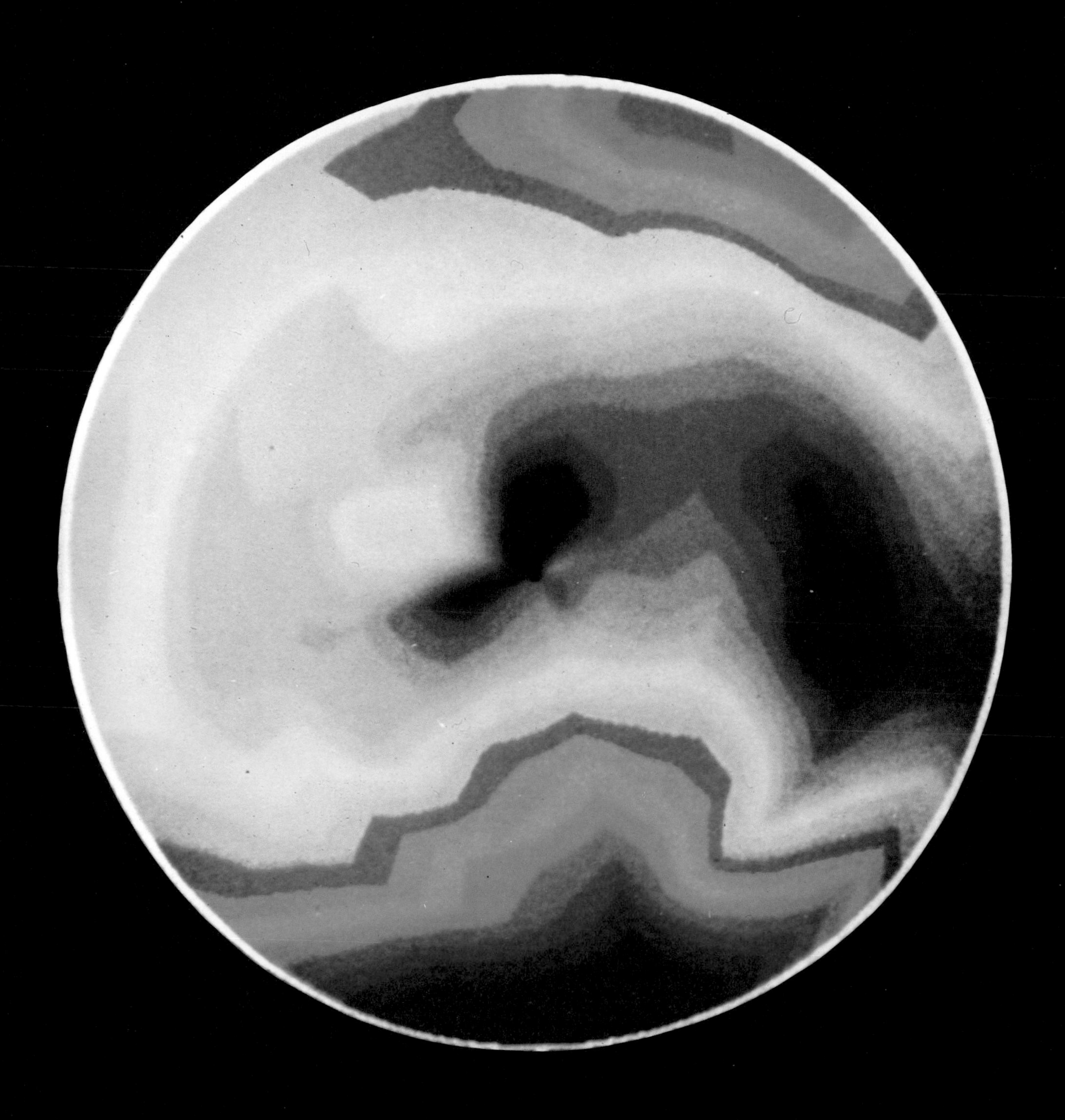

Ever since astronomers realized that Venus' surface was permanently hidden from their gaze by a dense mantle of cloud, they have been making inspired guesses as to what was down there. One of the most popular visions of Venus, which endured until only thirty years ago, was that the planet resembled the Earth during the Carboniferous Era. 'The humidity is about six times the average of that on the Earth, or about three times that in the Congo,' wrote Nobel Prize-winning chemist Svante Arrhenius in 1918. 'We must therefore conclude that everything on Venus is dripping wet . . . A very great part of the surface of Venus is no doubt covered with swamps, corresponding to those on Earth in which the coal deposits were formed . . .'

According to legend, Venus was born rising naked from the waves. Sandro Botticelli's beautiful painting of the event is in the Uffizi Gallery, Florence.

Less popular was the notion that Venus, closer to the Sun than the Earth, might be an arid dustbowl. Amongst the many speculations, that suggested by Fred Hoyle in 1955 was characteristically unique: 'We must now add the possibility that the clouds might consist of drops of oil . . . if Venus possesses oceans . . . the oceans may well be oceans of oil. Venus is probably endowed beyond the dreams of the richest Texas oil king.'

Nevertheless, it had to be admitted that the 'Carboniferous theory' had many more attractions than any of its rivals. Arrhenius concluded: '. . . some time, perhaps not before life on Earth has reverted to its simpler form or even become extinct, a flora and fauna will appear, and Venus will then indeed be the "Heavenly Queen" of Babylonian fame, not because of her radiant lustre alone, but as the dwelling place of the highest beings in our Solar System.'

Previous page: *Hot and cold zones in Venus' atmosphere show up in this false-colour 'photograph' taken by the US Pioneer Orbiter probe. Centred on the warm north pole (dark brown), the temperature falls towards the equator (deep blue).*

The association of Venus with goodness and beauty dies hard. In legend, Venus – the goddess of love – was daughter of the sky-god Uranus, and entered the world by rising naked from the sea. Her amorous conquests were many. It was said that she owed part

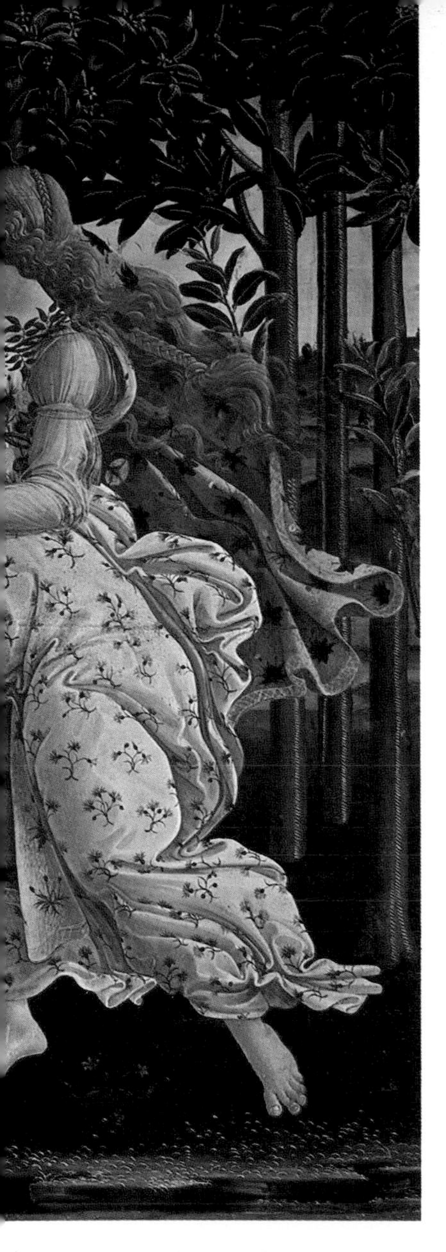

of her success to her magic girdle, which made its wearer irresistably lovely and desirable.

On first glance, there can be no doubt that the planet Venus was aptly named. Seen against an eggshell-blue twilight sky, Venus is by far the most beautiful of all the planets. Untwinkling and dazzling-white, it looks like a festival lantern hanging amongst the stars. As the planet which approaches Earth most closely – the two worlds are a mere 26 million miles (42 million km) apart at their nearest – Venus is also the brightest object in the night sky, apart from the Moon, and it can even cast a shadow on occasions. Needless to say, it is frequently mistaken for a flying saucer!

Like Mercury, Venus is only ever seen as a morning or evening star. But it lies further away from the Sun, and so it can draw away to quite considerable distances in the sky. At its greatest 'elongations', Venus can rise or set over four hours before or after the Sun, and sometimes, during the summer, you can see it in the sky at midnight. Venus is quite the opposite of elusive Mercury, but like its shy companion, Venus also goes through a cycle of phases, which depend upon where we see it in its orbit about the Sun. At its closest to us, it is a slender crescent. We can only see it as almost 'full' when it is far away on the opposite side of its orbit, and very nearly behind the Sun.

You can see Venus's phase – particularly when it's a nearby crescent – through a small telescope, or even with a well-supported pair of good binoculars. But the first person to see the phases of Venus, despite the small size and (as compared to today) poor quality of his telescopes, was Galileo. A methodical, persevering observer, Galileo was also a brilliant interpreter of what he saw. He realized at once that the phases Venus showed were proof that it circled the Sun, and not the Earth. True to the tradition of his time, he first announced his findings in an anagram to establish his priority as discoverer. 'Haec immatura a me iam frustra leguntur o.y.' translates to 'These things which are unready for disclosure are read by me.' The apparently meaningless letters 'o.y.' reveal that the sentence is an anagram, and that the reader should attempt to juggle the letters. Rearranged, it reads 'Cynthia figuras aemulatur Mater Amorum' – 'The phases of Cynthia (the Moon) are imitated by the Mother of Love.' This discovery, together with his observations of Jupiter's four largest moons, convinced Galileo that the Sun, and not the Earth, lay in the centre of the Solar System.

Although astronomers following Galileo also recorded Venus's phases, none of them managed to capture any markings on its disc. The task was so difficult that no one appears to have attempted any drawings of the planet until 1726, well over a century after Galileo first observed it. But despite Venus's unwillingness to co-operate over its markings, it continued to be useful – as it had for Galileo – in helping astronomers establish the true layout of the Solar System.

Captain James Cook led one of the many parties to observe a transit of Venus in 1769. His observations – from Tahiti – were somewhat hampered by the local inhabitants stealing everything from snuff-boxes to telescopes.

By the 1700s, there was no doubt about the fact that the planets circled the Sun. There was also no doubt about the *scale* of the Solar System. This had been established a century before by Johannes Kepler, who had shown that there was a link between each planet's 'year' – the time it takes to do one orbit – and its distance from the Sun. So while astronomers knew the *proportions* of the Solar System, they didn't know any of the actual distances. Just to find one distance would automatically lead to all the others.

One way of finding that magic distance would be to wait for an occasion when Venus makes a transit of the Sun – when it crosses the Sun's disc in silhouette. People observing the transit from different parts of the world would, due to perspective, register the crossing as happening over different parts of the Sun. Timing how long the crossing took, together with a little geometry, would give the distance to Venus – and the actual distances of *all* the planets from the Sun.

In the eighteenth century there were two transits, in 1763 and 1769. Among the many parties which set off to the corners of the Earth for the latter transit was an expedition to Tahiti led by Captain James Cook.

In the event, the observations of the transit – on 3 June – were not a great success. The same was true everywhere in the world, because of an effect noticed at the previous transit in 1861. Venus looked blurred, and this made timings very difficult. But what made Venus look so fuzzy? The eighteenth-century Russian astronomer Mikhail

Lomonosov put forward the correct explanation: 'The planet Venus is surrounded by a considerable atmosphere, equal to, if not greater than, that which envelops our earthly sphere.'

So here was the reason why Venus, tantalizingly close, never showed any well-defined markings. This didn't stop a number of eighteenth- and nineteenth-century observers reporting high mountains, but most of them just sketched vague shadings on the planet – the topmost layers of its clouds. From time to time, Venus does reveal more. Sometimes it's possible to see the whole of the unlit hemisphere of the planet glowing faintly – a phenomenon called the 'Ashen Light'. We often see this on the Moon – due to reflected Earthshine – but Venus has no moon, and the Earth is far too distant to

The clouds of Venus, photographed here in close-up from the Pioneer Orbiter probe, are made of drops of sulphuric acid. The streaks seen here are a sign of the rapid rotation of Venus' atmosphere – it circulates once around the planet in only four days.

light it. One of the most imaginative explanations for the effect came from an aristocratic German astronomer of the eighteenth and nineteenth centuries, Franz von Paula Gruithuisen. By establishing that there seemed to be an interval of 47 years between the glows, he concluded that they arose during festivals staged to celebrate the ascension of a new emperor to the throne of the planet. To give him credit, he did modify his ideas later on and suggested, rather less ambitiously, that they were forest fires caused by jungle being burned to create new farmland. Even his colleagues thought Gruithuisen was a little offbeam, and astronomers now believe that the ashen light may, possibly, be due to lightning.

Even in the mid-twentieth century, Venus's reputation as a planet of beauty and mystery remained virtually untarnished. No one knew what lay under those highly reflective clouds. The only certainties were that Venus was almost Earth's twin in size, that its atmosphere contained carbon dioxide, and, being closer to the Sun, it had a shorter 'year' of only 225 days.

The first flaws in the rosy picture began to emerge in the mid-1950s. In the early days of radio astronomy, some of the big dishes were pointed towards the planet. The signals they picked up suggested that beneath its clouds, Venus was searingly hot. But it was easy to dismiss the result as a mistake – an instrumental error, perhaps.

In the 1960s, however, it became apparent that Venus was more than just a little bit peculiar. The first peculiarity lay in the length of Venus's day, which until 1962 remained completely unknown. Then – as they had done in the case of Mercury – astronomers used radio telescopes to bounce radar waves off the surface of the planet and time its spin-rate. The radar waves penetrated the clouds, astronomers analysed the weak echo that returned – and found that Venus's day is 243 Earth days long, the longest in the Solar System and actually longer than the planet's year. What's more, Venus spins *backwards* – in the opposite direction to most of the other planets.

Space probes, too, were confirming Venus's growing reputation as a rogue planet. After a few early failures, Soviet space scientists rapidly established Venus as 'their' planet. But they didn't find it easy to explain why their first three perfectly functioning Venus landing probes – Veneras 4, 5 and 6 – stopped transmitting half-way down to the surface. Gradually, they arrived at the inescapable conclusion that the probes had been crushed out of existence by the immense pressure of Venus's atmosphere.

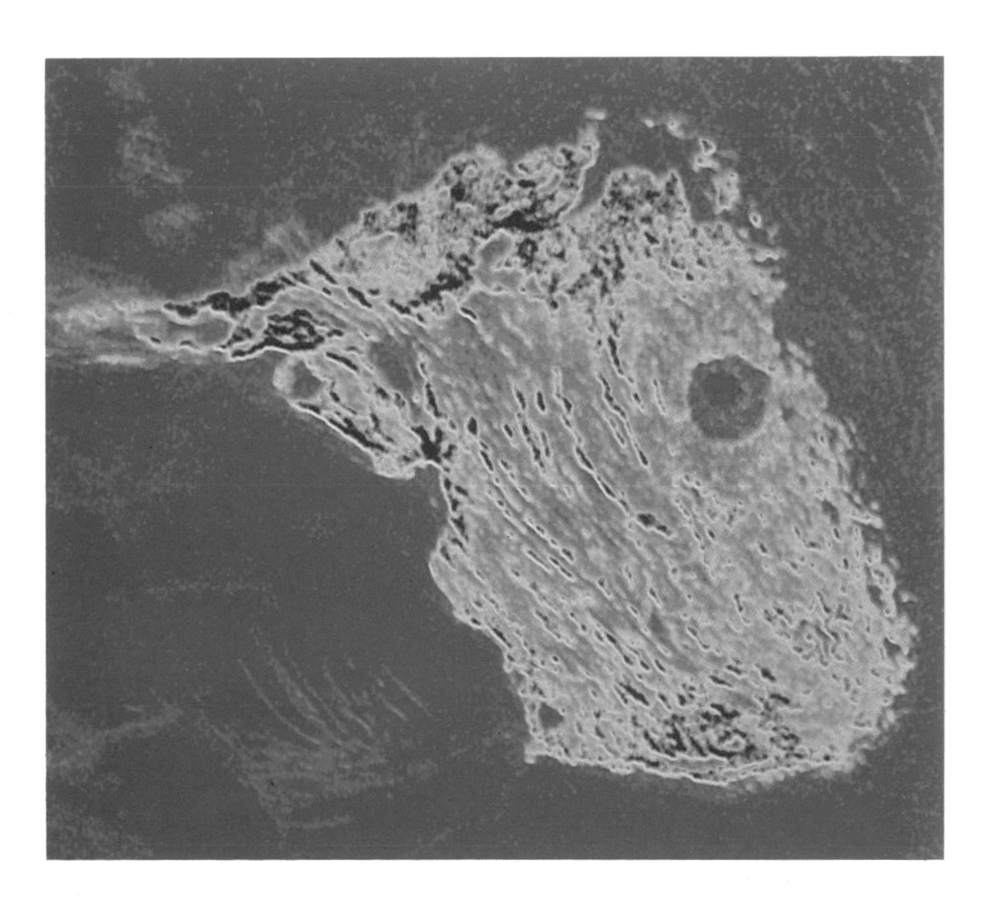

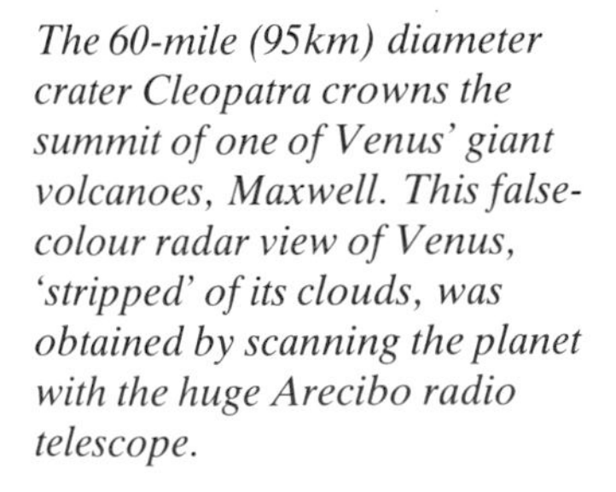

The 60-mile (95km) diameter crater Cleopatra crowns the summit of one of Venus' giant volcanoes, Maxwell. This false-colour radar view of Venus, 'stripped' of its clouds, was obtained by scanning the planet with the huge Arecibo radio telescope.

Later Soviet probes, looking like a cross between a bullet and a tank, finally made it to the surface. They confirmed everyone's worst fears. Venus's atmospheric pressure was ninety times that of the Earth – the same as that experienced by a submarine 500 fathoms (1000 metres) down in one of our oceans. The atmosphere itself was composed almost entirely of unbreathable carbon dioxide gas, in which floated clouds of sulphuric acid. And the temperature at the surface turned out to be around 470°C – the hottest in the Solar System – higher than the top setting on any domestic cooker.

It will be a long time – if it ever happens – before an astronaut stands on Venus's red hot surface to witness the desolation for him or herself. But camera-carrying Soviet probes have already shown us what to expect, while US craft have sampled the planet's surface and surrounding atmosphere. So thick is the atmosphere that the first Russian probes with cameras – Veneras 9 and 10 – took powerful lights with them to illuminate the anticipated gloom. To the surprise of the Soviet scientists, conditions weren't nearly as bad as expected, and the lights turned out to be unnecessary. 'It was

The sites where the two Russian Venera probes 13 and 14 landed were 560 miles (900km) apart, and each had a different kind of terrain. The rocks at the Venera 13 site (top), near the volcano Phoebe Regio, were small and fine grained. Venera 14 landed in a region covered in a platform of large, broken rocks. At both sites, the sky was orange.

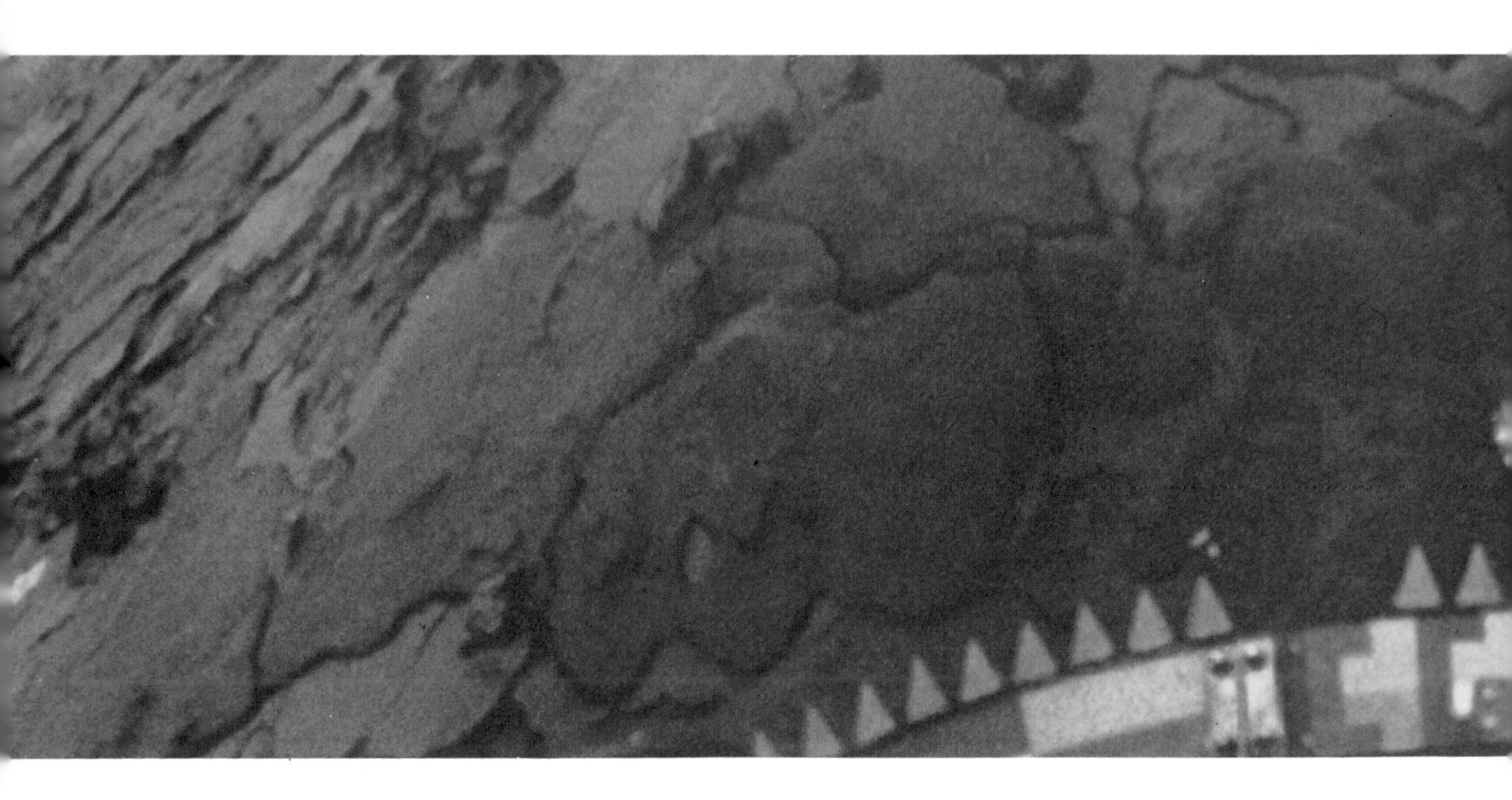

about as dark as we find on an overcast winter's day in Moscow,' claimed a spokesman.

Thanks to Veneras 9, 10, 13 and 14, we have had the chance to examine four different landscapes on Venus. All four show scenes of complete desolation, with boulders and flat slabs of rock stretching to the horizon. The colour cameras aboard Veneras 13 and 14 also reveal an unexpected twist: the rock strewn surface of the planet is bathed in the light of an orange sky!

The Veneras carried instruments for analysing the rocks and for interpreting local 'weather' conditions. Since one of the first probe findings about Venus's atmosphere was that it rotates about the planet in just four days – extremely fast in relation to Venus's 'day' – scientists expected that we would find tremendous winds at the surface. Surprisingly, conditions were very calm, with wind speeds of less than 1mph (2kph). And the rocks, expected to be sandblasted and eroded, turned out to be angular and

young looking. Most of all, they resembled the basalts we find in volcanic regions on Earth. Space scientists started to look at the planet afresh. In particular, they wanted to know whether Venus had volcanoes. And did the fresh, angular rocks mean that volcanic activity had been taking place relatively recently?

To find the answers to questions like these, astronomers have been busy for the past few years in trying to build up a global picture of Venus under its clouds. Since 1978, the US Pioneer Venus craft has been circling the planet and mapping its surface by radar. While its sister craft released a batch of secondary probes which transmitted data on Venus's clouds as they plunged towards the surface, the Pioneer Orbiter has been keeping the planet – and its atmosphere – under constant surveillance. Also in orbit around Venus are two Soviet probes, Veneras 15 and 16. They, too, have bounced radar waves off the planet's surface and returned pictures which show even finer detail.

But the most unexpected Venus mapper of all is based in a natural limestone hollow on the Caribbean island of Puerto Rico. Here, amongst green steamy jungle crusted with the orange flowers of Tulip trees, the world's largest radio telescope surveys Venus's arid plains. In its out-of-the-way location, close to the Earth's equator to get maximum sky coverage, the Arecibo radio telescope must surely have the most exotic site of any in the world. Puerto Rico's porous limestone rocks are pitted with natural craters where underground streams have caused subsidence. In one, scientists have strung this 1000-foot (305-metre) diameter wire mesh dish – a gigantic bucket to collect radio waves from the sky.

The 'business end' of the telescope – a moveable instrument platform – is suspended more than 500 feet (150 metres) above the dish. The journey to the platform – by walkway or cable car – is only for those with strong stomachs.

Instead of waiting passively for radio waves to come in from the sky, astronomers generate their own signal to bounce off Venus. Since the signal has to travel a minimum of 52 million miles (83 million km) there and back, it needs to be powerful.

The Pioneer Venus Orbiter mapped the planet's surface in detail by radar beneath the clouds. Apart from a few 'continents', Venus' surface consists largely of a plain interrupted in places by (probably active) volcanoes – like Beta Regio. With the exception of Maxwell, all the features on Venus have been given female names.

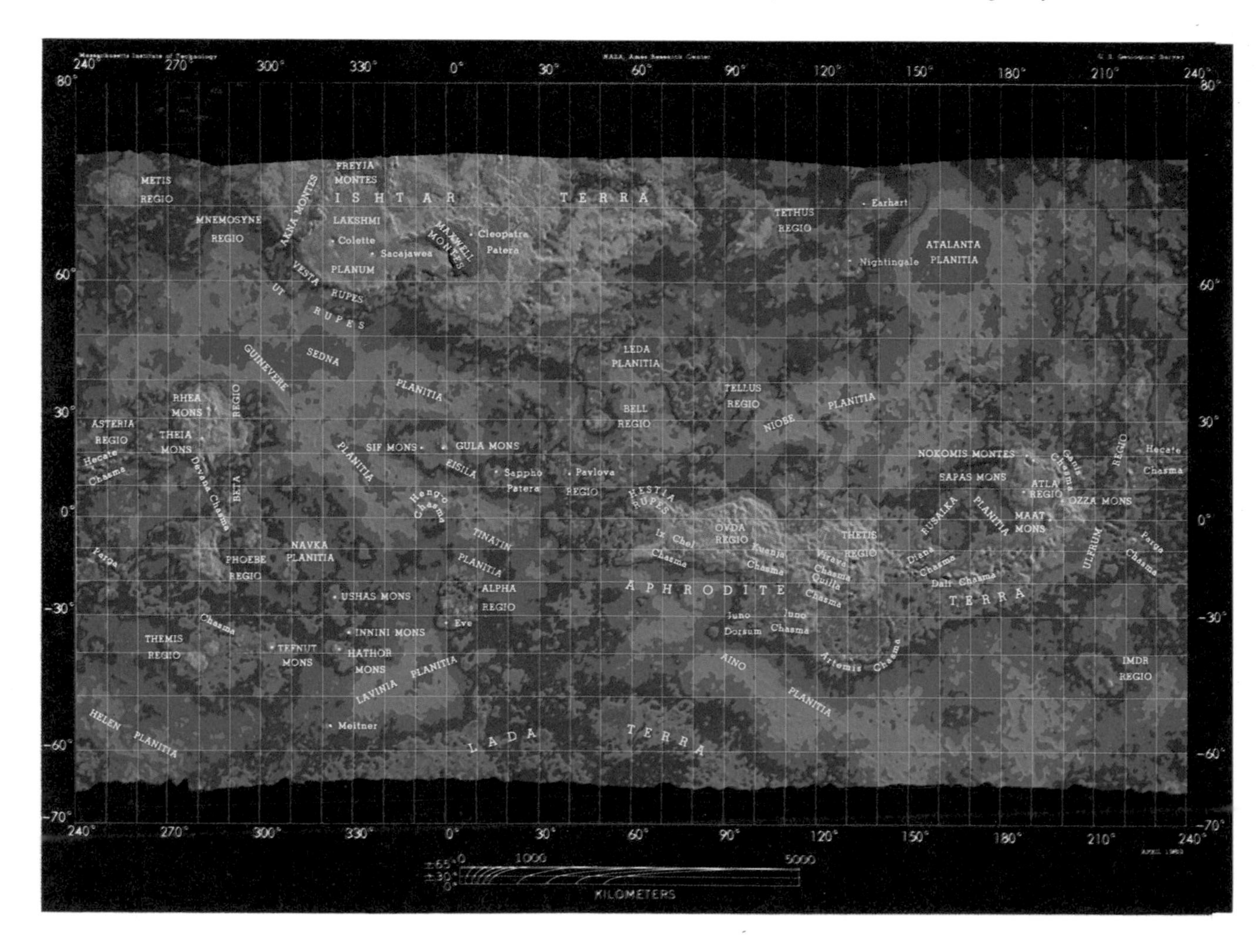

Pioneer Venus photographed a glow of ultraviolet light coming from the dark side of Venus. It may be related to the 'Ashen Light' sometimes seen from the Earth.

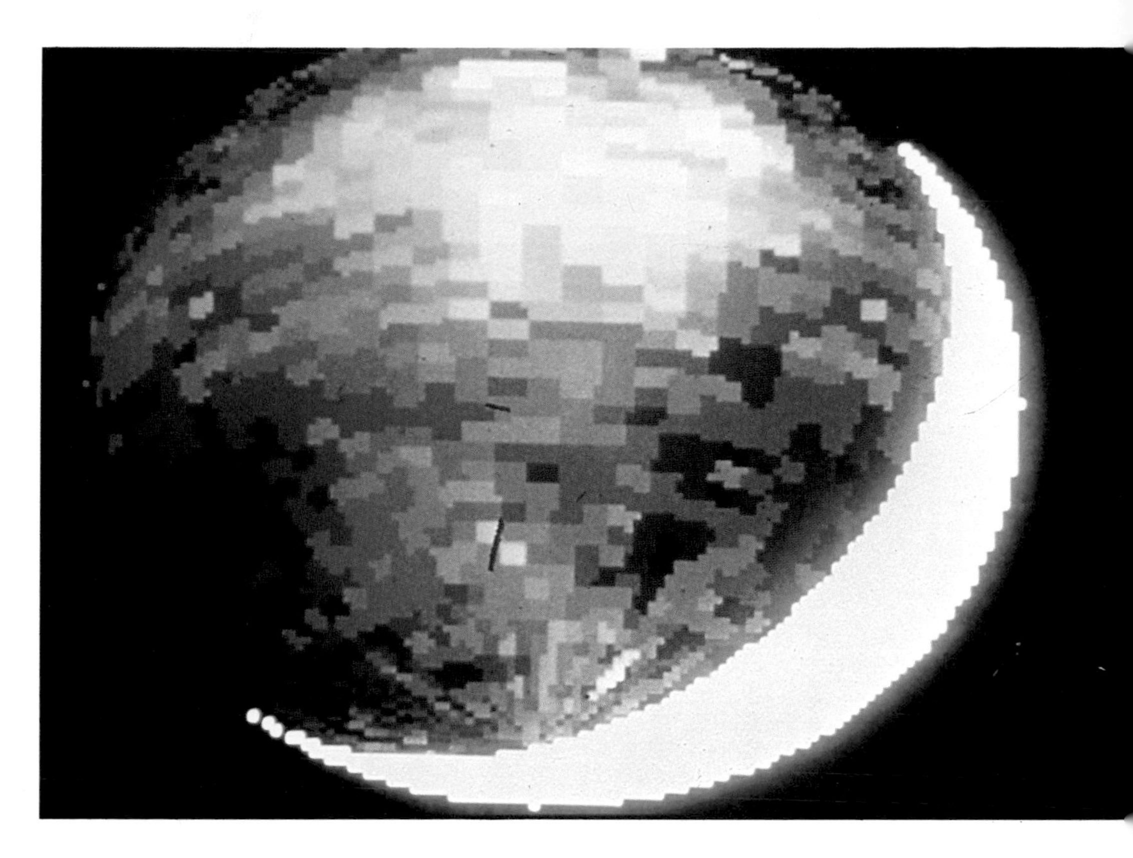

That's why people are banned from standing under the dish on mapping days in case they end up microwaved!

The radar maps of Venus from space probes and from the Arecibo telescope show a world whose surface is surprisingly different from ours. Our own planet – stripped of its oceans – would appear to have two distinct levels, the ocean floor and the continents. The continents, on average some 13 000 feet (4000 metres) above the ocean floor, cover about one-third of the Earth's surface. Venus, by contrast, is very much on one level. Most of its surface is covered by a flat, gently-rolling plain, interrupted in places by a few lower-lying basins. There are only three major highland regions, occupying in total just one-tenth the area of the whole planet: Aphrodite Terra, Ishtar Terra and Beta Regio. All except the first two features to be discovered have feminine names, an allusion to the goddess of love. And so Sappho, Leda and Helen rub shoulders with Guinevere, Colette and Lise Meitner.

Two of the upland regions – Aphrodite and Ishtar – are enormous plateaux rising sharply out of the planet-wide plain. Aphrodite is the larger, comparable in size with the continent of Africa, and it is situated just south of and parallel to Venus's equator. Its rough mountains rise to 4½ miles (7km) at the west of the region, while the 'continent' is sliced through by the huge Diana Chasma. Sinking to 2½ miles (4km) below the adjacent ridges, this enormous rift valley is 175 miles (280km) across at its widest – considerably larger than its counterparts on Earth, and comparable to the vast Valles Marineris system on Mars.

Ishtar, near Venus's north pole, is about the same size as Australia. It's made up of three main mountain ranges, of which the largest, Maxwell Montes (Maxwell Mountains) lies to the east. Maxwell itself is the highest point on Venus, and at 7 miles (11km) in height, it is taller than Mount Everest. In fact, cross-sections of Ishtar Terra bear some resemblance to the Earth's Himalayas.

The third upland region, Beta Regio, is different. Located at about 30°N, Beta consists of two cone-shaped mountains – Rhea Mons and Theia Mons – which look exactly like gigantic versions of 'shield' volcanoes such as Mauna Loa in Hawaii.

All the evidence now suggests that volcanoes have had a big part to play in shaping Venus; not only in the past, but recently, too. 'Venus is a very dynamic planet,' says Hal Masursky, a leading expert on terrestrial (Earthlike) planets at the US Geological Survey, Flagstaff. Masursky is one of the few Western scientists who has been deeply

The US Pioneer Orbiter probe has been circling Venus since 1978, studying the planet's atmosphere and mapping its surface by radar. In 1986 it will be used to investigate Halley's Comet when it is at its closest to the Sun.

involved in the Soviet exploration of Venus, as well as in the US effort. Because of his considerable knowledge of the geology of planetary surfaces, the Soviets ask him to advise them on the most interesting landing sites. In return, he gets their latest data before it is officially published.

At the moment, he is enthusing over the Venera 15 and 16 radar maps of the planet. 'They're fantastic. They show volcanic and tectonic belts like nothing else in the Solar System. The closest analogy would be the Jura when they were being formed.'

On close examination, nearly all parts of Venus show some signs of volcanic activity. The flat plains seem to be covered with lava flows. The 'Scorpion's Tail' – the eastern part of Aphrodite – is deeply rifted and is probably still volcanically active. Beta Regio 'is enormously larger than Hawaii,' says Masursky, 'and possibly bigger in volume than Olympus Mons on Mars.'

The latest Soviet maps reveal dozens of shallow, circular features – 'torcs' – whose irregular walls scarcely rise above the surrounding terrain. At first, they look like the

A Soviet Venus lander craft, of the type dropped by the Vega mission in 1985. Although the craft may look quite the opposite of high-tech, appearances are deceptive; it is deliberately designed to withstand pressures of 90 Earth-atmospheres and temperatures close to 500°C.

result of meteorite impacts, like the craters on the Moon or Mercury, but on examining the torcs carefully, Masursky has concluded that only one is definitely the result of an impact. The others are more uncertain. To find out exactly what they are will require mapping more detailed than that carried out so far. But Masursky's money is on vulcanism again – 'they may turn out to be old, much eroded volcanic centres.'

Whether or not Venus had a volcanic past, it is almost certainly volcanically active now. When the US probe Pioneer Venus arrived at the planet in 1978, teamleader Larry Esposito was astonished to register extremely high concentrations of sulphur dioxide gas and sulphuric acid particles above Venus's clouds. He continued to monitor the levels, and found that by 1983 they had dropped by 90 per cent. Esposito proposed that the source of the sulphur dioxide was a colossal volcanic eruption which, following Spode's Law (the astronomical form of Sod's Law), took place only months before the probe arrived at the planet. The eruption, if it happened, must have been truly enormous. Even the recent outburst of El Chichon in Mexico produced only one-tenth as much as gas.

Which volcano erupted? Many lines of evidence point to the region around Beta Regio, which was recently mapped in very fine detail by Don Campbell and his colleagues with the Arecibo radio telescope. His results reveal a vast rift, together with smaller 'volcanic' features which seem young, rough and uneroded. Pioneer Venus, still circling the planet, has been picking up bursts of radiation from the region. The most likely explanation is that these are flashes of lightning taking place above the erupting volcanoes, just as we find on the Earth.

While Venus certainly appears to be an active, dynamic planet, it is basically very different from our own. As we shall see in chapter four, it's thought that in the Earth's case, heat currents rising from deep in its interior push around the separate 'plates' which make up Earth's crust. Venus appears to have no plates. The radar maps show no telltale rifts and ridges on its surface; no characteristic jigsaw-puzzle pattern around its few 'continents'. From the size of the volcanoes in Beta Regio, it seems likely that they have grown so large because they have been sitting on an underlying hot plume since the plate was formed. No volcano on the Earth is allowed to become so big; the shifting crust takes care of that.

So while Venus has a hot interior, its surface has turned out to be very different from that of its sister-planet. It's a puzzling difference to explain, because Earth and Venus are so much each other's twin, in size and mass. Perhaps more clues will come from the US Venus Radar Mapper probe (VRM), due to be launched in 1988. This orbiting probe will 'see' details on Venus' surface only as wide as the ribbon of the River Thames snaking through London. And they hope to follow it at a later stage with a probe which will snatch a sample of Venus-rock and return with it to Earth. One scientist has remarked that the technology required for the return-leg of this mission would be very similar to that involved in launching a missile from a submarine in the depths of the ocean!

In early 1985, Soviet spacecraft were on their way to Venus again, in the form of the two Vega probes. The main part of each craft is to fly on to Halley's Comet in 1986 – at a safe distance – but, *en route*, the Vegas each dropped an instrument-laden balloon down through Venus's atmosphere, to float at a height of 34 miles (54km) above Venus's surface. There has been international collaboration on this project, with the Soviet balloons being tracked by the telescopes of the US Deep Space Tracking Network at Goldstone, California. Very soon there should be some new results to hand concerning Venus's only-too-ample atmosphere.

But why is Venus's atmosphere so ample? And why do Earth and Venus differ so greatly when, on the face of it, they should be so similar? According to many planetary scientists, Venus and the Earth started life very similarly. The young planets, still restless and hot, seethed with volcanic activity. The volcanoes belched carbon dioxide and water vapour into the air. On Earth, further from the Sun, it was relatively cool; the water vapour ultimately condensed as rain and covered the planet with great oceans. The carbon dioxide, too, slowly vanished from the atmosphere. Some scientists believe that it was all dissolved in the seas; others think it was removed by a new upstart on the scene – life.

Venus, just 30 per cent closer to the Sun, was always just too hot. Perhaps some of the water vapour did condense, and the Pioneer Venus results on the atmosphere indeed suggest that it did. There may, for a while, have been rivers and streams – something which planetologists are currently searching for now. But there was never enough water to dissolve the carbon dioxide, and so it built up in the atmosphere with every eruption.

Carbon dioxide has the same effect as a pane of glass in a greenhouse: while it lets light through, it doesn't allow heat to escape, and so the temperature of the planet increased. It wasn't long before what little water there was had been boiled away. Eventually, it vanished into space as separate atoms of hydrogen and oxygen. Thanks to a 'runaway greenhouse effect', the surface temperature of Venus has now risen to 470°C.

Ironically, the continuing volcanic eruptions may be responsible for keeping the planet cooler than it would otherwise be. The sulphuric acid clouds, almost certainly formed from sulphuric dioxide gas shot into the atmosphere during eruptions, reflect back four-fifths of the light they receive from the Sun. Without this screen, Venus might be an inferno.

There is a lesson here for us. It lies in the fact that we, too, are increasing the levels of carbon dioxide in our atmosphere at the moment. By burning fuels like coal and oil – the fossilized remains of carbon-rich plants which flourished hundreds of millions of years ago – we are releasing vast quantities of the gas into our air every second. Two independent studies conducted in the US reveal that our planet is warming up as a result.

The situation is so complex that it's very difficult to be able to say how much, and how quickly. There are many other factors to be taken into account, not the least of which is the Earth's tendency towards ice ages. But there is a clear message that we are putting our planet at risk if we continue to burn fossil fuels.

Detractors may argue that space probe research on planets is money ill-spent. But do they realize how much it tells us about our own planet? If, in the long run, it enables us to safeguard what we have, then every penny will have been worthwhile.

CHAPTER FOUR

THE EARTH

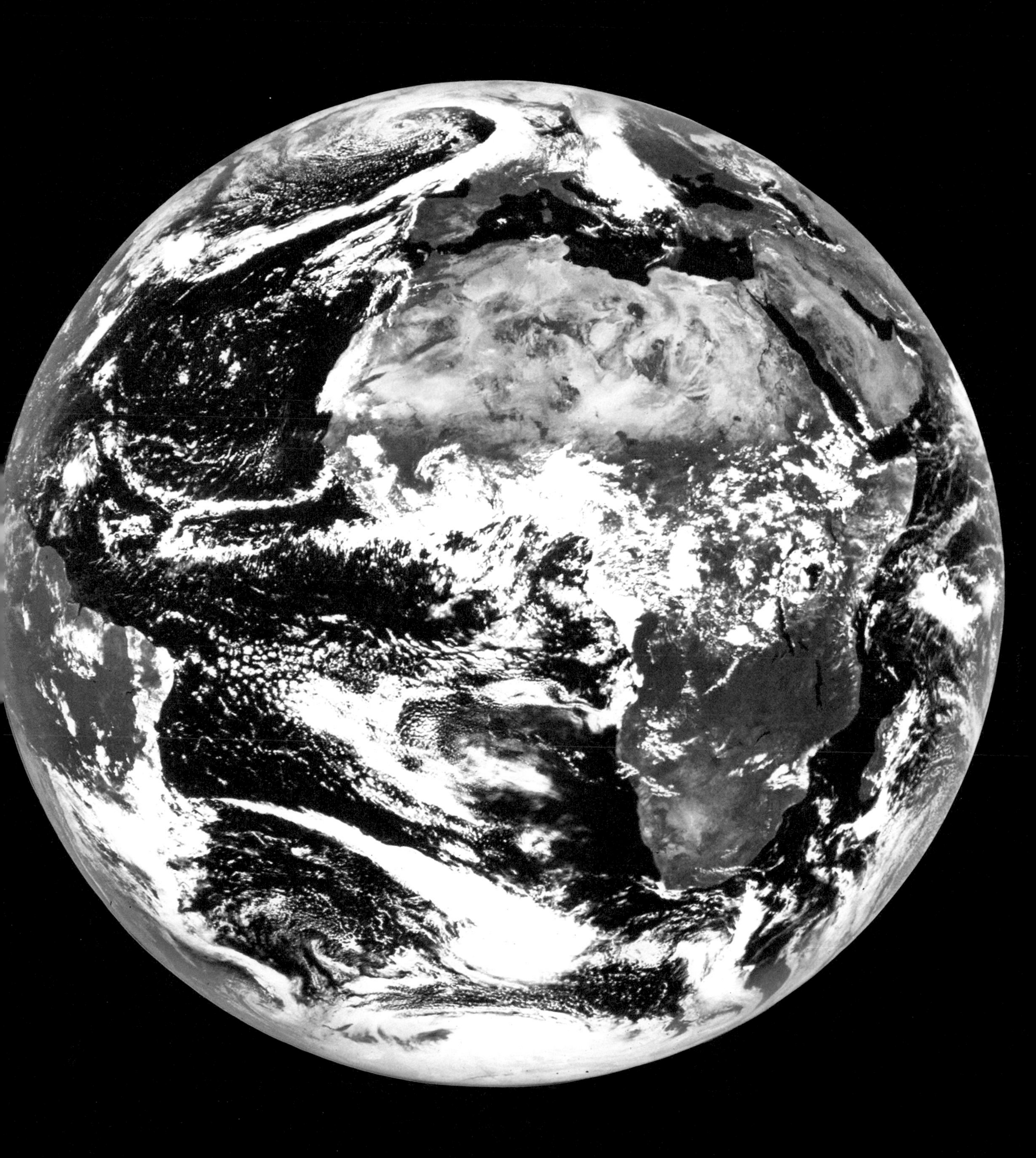

This early eighteenth-century orrery shows the Earth as dominating the inner Solar System. In comparison, the Sun and the inner planets Mercury and Venus appear insignificant.

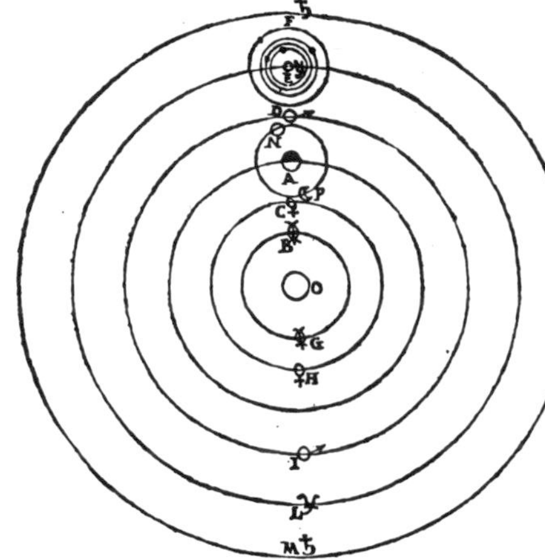

Galileo's map of the Solar System shows the Earth in its rightful place – as just another planet circling the Sun. His map also shows the orbit of the Moon, and those of the moons of Jupiter.

At first sight, the Earth beneath our feet seems to have little in common with the 'wandering stars' that the Greeks called the 'planets'. The Earth is large and solid, and apparently immobile under the wheeling skies above. Although some Greek philosophers did begin to challenge the Earth-centred view – Aristarchus, for example, believed that the Earth moves around the Sun – such views were not very popular. Right up to the sixteenth century, Aristotle's philosophy and the Bible were twin authorities that kept all right-thinking people in line with the idea that the firm Earth beneath our feet is fundamentally different from those distant lanterns in the heavens.

In 1543, however, Nicolas Copernicus started the great revolution that changed our concept of the Universe. He dethroned the Earth, and put the Sun at the centre of the system of planets – with the Earth itself as merely one of the six planets then known. At the beginning of the next century, Kepler and Galileo proved that Copernicus was correct. Galileo not only showed that the planets go around the Sun. When he saw the changing phase of Venus, he was, for the first time, making out the globe of another planet. In the next few decades, he and his successors were able to discern the other planets as worlds in their own right.

Even so, it took a long time for people to come to terms with the idea that the Earth is merely one of half-a-dozen worlds. Part of the problem was that astronomers did not know the true scale of the solar system. Kepler had provided an accurate model, but no one knew the actual distances involved. The Greek astronomers had tried to measure the distance between the Earth and the Sun, and had come up with a value of about five million miles (8 million km) – only one-twentieth of the correct value – and Kepler had guessed a value only slightly larger. As a result, the seventeenth-century astronomers were wildly wrong when they tried to gauge the size of the other worlds of the solar system. The Sun, they believed, was only a few times larger than the Earth; and the other planets were insignificant globes when compared to our world.

The man who changed all this was Giovanni Domenico Cassini, a great Italian astronomer who in 1669 moved to the newly-founded Paris Observatory. Cassini realized that he could find the distance to a planet – he chose Mars – by measuring its position very accurately at dusk, and then again at dawn. During that 12-hour period, the Earth's rotation had moved the Paris Observatory several thousand miles side-

Previous page: *This view of Earth, taken by the Meteosat weather satellite, shows why astronauts call it 'The Blue Planet'. Two-thirds of the Earth's surface is covered by water, unlike any of the other planets.*

As Voyager 1 departed into the depths of space, never to return, it took this last look back towards the Earth and Moon. The slender crescents are 7 million miles (11 million km) away.

ways. As a result, Mars should appear to have moved in the opposite direction against the background of distant stars – just as a finger held at arm's length seems to jump backwards and forwards as you observe it with one eye, and then the other.

Cassini found that Mars's motion was much smaller than he expected – meaning that the Solar System was considerably larger than anyone had believed. Although his method for plumbing the depths of the Solar System was later superseded by more accurate means, Cassini had broken important new ground. The Earth, he realized, was a relatively small planet. Jupiter and Saturn are much larger than our world, and Venus is a close twin.

As this realization sank in, scientists, philosophers, and popular writers began to draw comparisons between our planet and the others – an approach that we would today call 'comparative planetology'. Three centuries ago, astronomers were faced with the problem that their telescopes could reveal little about the other worlds, apart from their sizes. So they simply assumed that the other worlds were similar to the Earth.

Scientists and writers alike peopled the worlds of the Solar System with animals and intelligent beings. Well, why not? If the Earth is a typical planet, then surely life must be common. Their religious upbringing also played a role. God would hardly have bothered to make the other five planets, unless they were a home for intelligent beings, just as he made the Earth as Man's home. Christiaan Huygens, the great astronomer who discovered the rings of Saturn, argued that each of the planets must have a race of intelligent beings. He even deduced that the inhabitants of Jupiter and Saturn must be great sailors, 'having so many Moons to direct their Course'. A century later, Sir William Herschel – the discoverer of Uranus – was happily peopling all the worlds of the Solar System. According to Herschel, even the Sun was Earthlike, under a flaming atmosphere, and 'we need not hesitate to admit that the Sun is richly stored with inhabitants'!

As astronomers learnt more about the other planets, they found that they were far from being carbon copies of the Earth. In recent decades, space probes have shown the planets as they really are: barren, desolate and hostile. The Solar System has four great gas-and-liquid planets, which bear no family resemblance to the Earth. Even our neighbours in space seem to have more differences from the Earth than similarities. The rocky surfaces of Mercury and Mars are scarred with craters. The atmospheres of Venus and Mars are made of suffocating carbon dioxide gas; and their skies appear orange-red in colour. These planets are bone-dry. Their water has either been boiled away, or frozen solid. The barren plains of our neighbour planets seem to be devoid of life.

In contrast, the Earth is a living world. Figuratively, it is alive with flowing water, and with electric blue skies. And, of course, it abounds with living organisms. Great rain forests cover enormous tracts of the land around the Earth's equator, and trees, grass and crops stretch up through more temperate latitudes towards the polar regions. Both the sea and the land teem with animal life.

A space traveller visiting the Solar System would notice at once that the Earth is unique. Among the four inner worlds, she would see that the third from the Sun has a large moon – so large that she might call the planet plus moon a 'double planet'. More remarkably, the third planet shines with a distinct blue tinge. The other rocky worlds – and the third planet's moon – range through various shades of red, from Venus's very pale yellow to Mars' strong ruddy glow.

Drawing closer to the unique blue planet, our traveller finds that its colour is due to immense quantities of liquid, filling the low-lying regions that cover two-thirds of its surface. This liquid is water, for the blue planet lies at just the correct distance from the Sun for water neither to boil nor freeze. Even so, the space traveller sees that the planet's poles are covered with thick layers of ice, that expand and contract with the seasons.

Now she finds a second unique fact about this planet. Its atmosphere is not made of carbon dioxide, the gas that shrouds most of the other planets. Most of the gas is nitrogen, which is fairly common on other worlds. But the surprise comes with the other gas: it is oxygen, one of the most reactive gases in the Universe. Take some of the most common substances in the Cosmos – hydrogen, methane, or meteorites rich in carbon or iron – and put them in an atmosphere of oxygen, and you will find that the oxygen destroys them, turning them into water, carbon dioxide or rust. How did such a reactive gas arise on a planet?

A third discovery solves the puzzle. The traveller spots large areas of green, infesting the planet's surface, and largely hiding the natural redness of sand and rock. The greenery comes and goes with the seasons: it is living. These living plants mop up the little carbon dioxide that exists in the atmosphere, and with the help of the Sun's light they split it up. A plant uses the carbon from the gas to build itself up into large and beautiful structures; it rejects the oxygen atoms from the carbon dioxide as waste.

Finally, the space traveller finds another kind of living thing – a destructive organism. These creatures actually live on the dangerous oxygen gas in the atmosphere, and get their energy from ingesting the tissues of the green plants, or of each other.

The space traveller's view reveals that the Earth has several unique features. But our planet also has a deep family likeness to Mars and Venus. It is made of the same kind of rocks; and the interiors of the twin planets, Venus and Earth, are probably very similar. All three planets have volcanoes, even if those of Mars are now dead. The crust of Venus seems to have stretched and contracted to build up mountains, in a weak imitation of the processes that have thrown up the great mountain ranges of the Earth.

In the late twentieth century, the scientists studying the Earth – geologists and geophysicists – are coming together with astronomers in the new science of comparative planetology. A comparative planetologist seeks to understand all the worlds of the Solar System, by looking at the ways in which the planets are similar and the ways that they differ. Geologists and geophysicists have been studying the Earth intensely for over a century, and a comparative planetologist can use that knowledge to help him comprehend, for example, the volcanoes of Venus or processes of erosion on Mars. Conversely, our new views of the other planets are casting light on the workings of Earth.

The key to the Earth's uniqueness is its liquid water. Water has played many roles in shaping the Earth, literally as well as figuratively. As water flows downhill in streams and rivers, it wears away the ground below. Over the millions of years of geological time, rivers can smooth down great mountain ranges, until nothing is left but a low-lying soggy plain. As the mountains are washed away, grain by rocky grain, their matter piles up in deep layers on the ocean floors. The weight of newly-deposited matter squeezes the sediment below, and welds the fragments together as solid rock.

Erosion has entirely rubbed away the oldest rocks of the Earth's surface, wiping clean the record of our planet's formation. We could learn little of its earliest history were it not for clues from other parts of the Solar System. Meteorites can stray from their home in the asteroid belt beyond Mars, and fall to Earth, bearing tell-tale signs that the Solar System was born 4½ billion years ago. That, then, must be the age of our planet. Astronomers can tell us that the rocky worlds were built up from smaller hunks of rock, and that a final hailstorm of rocks blasted out the craters that we see jumbled all over the Moon and Mercury. These two small worlds have been dead since that time, and their surfaces still reflect the way that the early Earth must have looked.

There is one exception. The Earth had active volcanoes, which belched forth fire, brimstone – and, most important, gases. Our planet, unlike the Moon or Mercury, had sufficient gravity to hold on to these gases. They built up around the Earth, forming the primeval atmosphere. The gases coming from volcanoes today can give us a fair idea of what the Earth's early atmosphere was like. Most of a volcano's emissions consist of steam and carbon dioxide; mixed in is a smaller portion of nitrogen and more noxious gases, such as sulphur dioxide. The steam condensed into huge clouds that blanketed the Earth. Then the rains came. The flood lasted not for forty days, but for millions of years. The skies eventually cleared, on a sight new in the Solar System. The Sun shone on a planet that was covered with great oceans of water, filling the lowest-lying plains between bare, cratered continents. The oceans dissolved most of the gases from the early atmosphere, leaving an air that was relatively thin – compared to the thick atmosphere that dry Venus was accumulating – and made largely of carbon dioxide and nitrogen.

Earth is unique in the Solar System for its moistness and for the life it possesses in abundance. This rain forest epitomises the richness of our planet – green, damp, and teeming with living things.

The rains had also washed down something else: the basic molecules of life. Most scientists now believe that life originated in the thick atmosphere that originally enveloped the Earth. This gas was seared by bolts of lightning and shocked by infalling meteorites; the Sun's intense ultraviolet radiation irradiated its upper regions, and hot volcanic rocks supplied heat radiation from below. The atoms within the simple molecules of carbon dioxide, water and nitrogen broke apart, and came together in new formations, in a microscopic barndance of changing partners. In the process, some molecules grew in size. Instead of containing just two or three atoms, they grew to include five, ten, twenty atoms. The carbon atoms, in particular, tended to cling together. As they formed into a chain other atoms clung to the sides, to build up some of the complex chemicals – such as amino acids – that form the basis for living cells.

For years such ideas were speculation. But in 1953, a chemist, Stanley Miller, decided to see if it would work. He made his own miniature version of the Earth's primeval atmosphere: a glass flask filled with suitable gases, and a high-voltage spark as lightning. After a few days, the sides and the bottom of the flask were coated with a sticky yellow substance. Miller analysed the yellow gunge, and found it was indeed made of the kind of carbon chemicals that are found in living organisms.

The Earth is constantly rebuilding itself from inside by volcanic eruptions. In this photograph of the 1983 eruption on Hawaii's Kilauea, a stream of solidifying lava can be seen snaking down the mountainside.

Miller's experiment surprised many scientists. But it has now become clear that carbon atoms will join together whenever given the chance. Saturn's largest moon, Titan, for example, has an atmosphere thick with the kind of gunge that Miller created in his flask. Here we have another example of a stage of the Earth's development, fossilized for us to inspect – when we can send a space probe to delve into the orange clouds of Titan.

On Earth, the organic molecules were washed into the seas, and, somehow, they came together to form the first living cells. That 'somehow' bridges one of the largest gaps in human understanding. Even the simplest living cell is an incredibly complicated system. Extremely long molecules, called DNA, carry a special code that tells the cell how to work, how to grow, and how to replicate – in the simplest possible way, just by splitting in two. Other parts of the cell obtain energy; others repair damage. The cell combines the physical complexities of an automatic car-assembly line with the information handling of a computer. And all this came about by a suitable conjunction of chemicals dissolved in the seas of the early Earth – the 'primeval soup'.

Once the first living cells had formed, they could replicate over and over again, living off the surrounding chemicals. But eventually these chemicals would run out. Some cells saved themselves by acquiring a green chemical called chlorophyll that could trap energy from sunlight, and used this power to break down the carbon dioxide gas dissolved in the seas. From these cells grew all the plants we find in the seas and on land today – from plankton to trees, from seaweed to roses.

Plant power was able to change our planet. Over hundreds of millions of years, the plants removed the carbon dioxide from the Earth's atmosphere, and replaced it with their waste product, oxygen. The oxygen built up until it replaced carbon dioxide as the second-most abundant gas in air, after the rather inactive gas nitrogen.

Other cells developed to complement the plants. They breathed in the oxygen, and used it for 'burning' fuel within the cell, to produce energy and liberate carbon dioxide, which was then available for the plants to breathe. But these animal cells needed a source of fuel (or food) to burn. They got this by eating plants, or other animal cells. Eventually these single-celled animals evolved into the wide diversity of animal life we now know, in the sea and on land. And from the land-based animals, the process of evolution eventually produced human beings.

Planet Earth is unique in supporting life; but it also has a life of its own. The Earth's crust is continuously in motion, shifting the continents about the globe and destroying rocks that once formed its surface. Some of the other worlds in the Solar System show cracks where sections of the surface have begun to move, and mountains apparently pushed up in the process, but only on the Earth does the process go full circle. Like the treads on an airport 'travelator', rocks appear at one point on Earth, travel across for hundreds or thousands of miles, and then disappear back into the Earth's interior. Earth-scientists call the process plate tectonics: the 'plates' are the moving sections of the Earth's crust, and 'tectonics' means the processes that build mountains. Plate tectonics is a comparatively new theory; but it explains many mysteries of geology, including continental drift.

In a less dramatic way, the Earth also builds new land along its ocean floors. New ocean floor created along the Mid-Atlantic Ridge is pushing apart America and Europe at a rate of half an inch (1.2cm) a year.

In 1912, a German scientist, Alfred Wegener, suggested that the continents must have moved around the Earth. The idea seems obvious when we look at the South Atlantic, and see that the eastern bulge of South America fits snugly into hollow on the west side of Africa. In fact, the east side of North America also matches up to the seaboard of Europe. These jigsaw-puzzle fits suggests that the Atlantic once did not exist: the Americas were joined to the Old World as a single land mass.

Further evidence for continental drift came in only slowly over the decades. The world's leading geologists refused to believe that continents could move. Even Australia, the smallest continent, weighs almost a million million million tonnes, and how could it push through the solid rocks of the sea floor?

The answer began to emerge when Earth-scientists started to study the rocks that lie beneath the oceans. In the 1950s they came across several surprises in the Atlantic. First, there is an immense underwater mountain ridge that runs exactly down the centre of the Atlantic, its highest peaks rising above sea level as islands like Iceland and Tristan de Cunha. Secondly, the rocks of the ocean floor are remarkably young when compared to the Earth's age – less than 200 million years old. The final discovery was the real eye opener. On either side of the mid-ocean ridge, the sea floor bore stripes of magnetized rocks – and the patterns matched precisely across the central ridge as if it were a mirror.

Two geophysicists at Cambridge, Fred Vine and Drummond Matthews, took these clues, and worked out their implications. They reasoned that new rocks must be emerging at the mid-ocean ridge, and then travelling away to either side as new sections of the sea floor. Each new section bore a magnetic pattern, imprinted as it was created by the Earth's varying magnetic field – causing the magnetic pattern to be identical on each side of the ridge. As new sections of the sea floor are created by the centre, so the edges of the ocean – marked by Africa and South America – must move apart. So this is the reason for continental drift. The continents are not like great liners, steaming through the adjacent sea floor. The sections of sea floor are moving, and carrying the continents with them.

Geologists have now mapped the moving plates that form the Earth's surface. Large plates carry the continents of America, Africa, Antarctica, Eurasia, and Australia, while another big plate forms the floor of the Pacific Ocean. Between these plates are tucked half a dozen smaller ones.

Down the centre of the South Atlantic Ocean, the African and American plates are drawing apart, and adding new material to their trailing edges at the mid-ocean ridge. What happens at the other side of South America? The edge of the plate here coincides with the edge of the continent, and the massive moving block is running full-tilt into the rocks that lie beneath the South Pacific. South America rides up over the edge of the Nazca plate, and forces its rocks to dive down underneath the continent. In this subduction zone, the sinking rocks grind over one another, and snap apart deep below the Earth's surface, until eventually they melt in the Earth's hot interior. The process results in the earthquakes that regularly shake the inhabitants of Chile and Peru. These rocks carry a load of wet sediments from the ocean floor, and when they eventually melt the Earth's heat boils the water, and it rises in a mixture of superheated steam and molten rock, to break through the Earth's crust and form the magnificent range of volcanoes that we call the Andes.

About fifty million years ago, the plate carrying India, then a small continent in its own right, careered into Asia. Continents always ride high, and cannot duck under one another. So the result was an immense head-on smash. The Earth's crust rose up and buckled to form our greatest mountain range, the Himalayas, rising to our planet's highest point, the 29 000 feet (8800m) Mount Everest.

America's pleasant western state of California is a place where two plates adjoin, along the line of the San Andreas Fault. The land to the west of the fault is actually an interloper on continental America: it is a bit of land attached to the mainly underwater Pacific plate. The two plates are rubbing together, as the western part of California moves north at the rate of two-and-a-half inches (6cm) per year. Unfortunately for those who live in California this doesn't happen smoothly. The rocks on either side of the fault will bind together for years, or decades; then suddenly break apart and slip sideways. The resulting earthquake can be devastating, as San Francisco found in 1906.

Although plate tectonics is a wonderfully elegant theory, explaining so much of the Earth's geology, there was no direct proof that the Earth's plates are moving. In a way, that was not surprising. Someone in Europe keeping an eye on America should be seeing the continent move away at a speed of less than one inch per year – and he would have to measure that speed from thousands of miles away.

British scientist David Smith was not put off by the problem. When he started working in NASA's Goddard Space Flight Center, near Washington DC, Smith realized that he could measure the continents' motion, if he had enough time. NASA and American astronomers were already using equipment that could give the distance between two planets to a precision of a few inches. All that Smith needed was to keep on making the same measurements from the same locations, and see how the distance between them changed as time went by. Thus was born the Crustal Geodynamics Project: an audacious attempt to lay measuring rods around the Earth, using two different space age techniques.

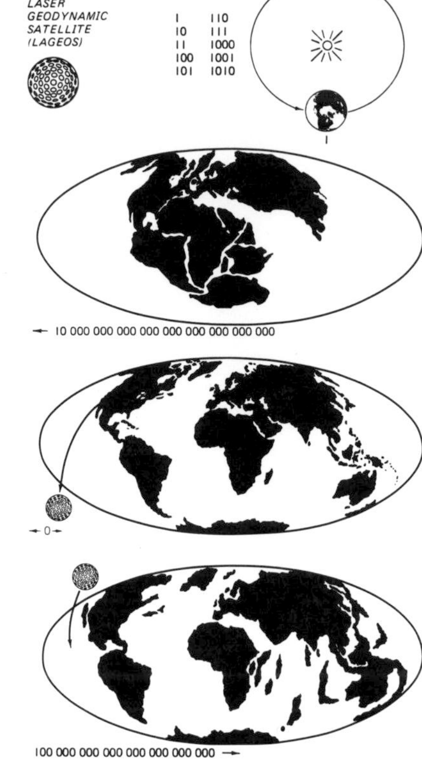

The Earth is the only planet in the Solar System to have a constantly-shifting crust. The continents are situated on moving 'plates' which, over millions of years, carry them around the planet. This reproduction of the plaque on the LAGEOS spacecraft shows how the Earth's continents were arranged about 300 million years in the past; how they look at the present time; and how they will appear 10 million years in the future.

Every clear night, and even during the day, scientists at the Royal Greenwich Observatory point a laser at the sky. Their target is a small satellite, studded with reflectors. The satellite reflects the laser light back to the observatory; and a telescope attached to the laser picks up the faint reflected flash of light. Electronic circuits calculate how long the flight has taken to get to the satellite and back, and so take a range on the satellite. Meanwhile, David Smith's lasers, across the Atlantic, are ranging the same satellite. Their repeated measurements will tell them the satellite's exact orbit – and the precise distance from the Royal Observatory to the laser station near Washington.

Radio astronomers do it a different way. Since the late 1960s, they have made a habit of 'looking' at the same thing in the sky – often a distant quasar – with two radio telescopes that are thousands of miles apart. What they are after is the finely-detailed view that this technique can show; but what also falls out of their calculations is the distance between the telescopes.

Smith uses both these methods to find distances across the globe. There are two dozen laser stations, and some twenty radio telescopes, that send in information to his project. So, if you want to know the precise distance from a radio telescope at Haystack, Massachusetts, to its counterpart at Onsala in Sweden, David Smith won't fob you off with 'about 5600km': in May 1984, the distance was precisely 559 971 455cm. The game is to measure the distance over a period of years, and see it gradually changing.

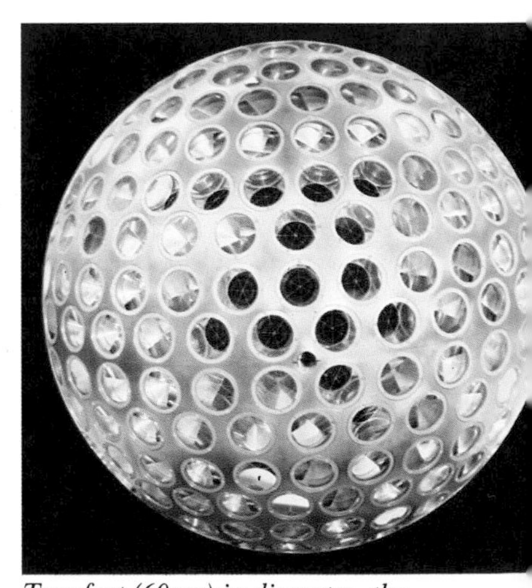

Two feet (60cm) in diameter, the LAGEOS satellite is studded with 426 reflectors to reflect back laser beams from Earth. This aluminium 'golfball' has enabled scientists to measure plate tectonics in action.

By 1984, Smith's team had amassed enough data to prove that the continents do move. The radio telescopes in Massachusetts and Sweden are gradually drifting apart, at a rate that shows the Atlantic Ocean is widening at the rate of half an inch per year. The laser stations around the Pacific show that the plates there are in much more rapid motion. Australia holds the record, steaming north-eastwards at almost three inches per year. As it overrides the Pacific plate, it pushes it down into some of the world's deepest ocean trenches – as far beneath sea level as Everest is above. The Crustal Geodynamics Project has not only proved continental drift – after all the decades of controversy – it has also shown that we can no longer think of the Earth like an unvarying school globe. American schoolchildren will learn not only that Europe is to the east, and Australia is to the south-west, but that the American continent is moving away from Europe and towards Australia. In future, the most accurate maps will tell us not only where our home town is, but how fast and in what direction it's moving over the Earth.

Modern Earth-sciences can tell us not only about the Earth's crust, but about its interior. For many decades, geophysicists have known that the Earth has a liquid core,

A beam of laser light is directed towards the LAGEOS satellite from the Mount Heleakala Observatory on Maui, Hawaii. By timing the return echo, astronomers can find the precise distance to the satellite and, in time, the rate at which the land is moving relative to it.

because the liquid causes a distinct 'shadow' in the earthquake waves coming through our planet. Now Don Anderson in California is setting up a system to probe the Earth's interior in fine detail. His technique is similar to that of X-ray machines in hospitals. Originally, doctors used X-rays just to cast a shadow of the internal organs, and the shadows of various structures inside the body obscured one another. Hospitals now have CAT-scan machines that X-ray the body from all directions, and build up a three-dimensional view of its interior. Anderson is starting to 'CAT-scan' the Earth, with the help of seismometers, the detectors that pick up earthquake waves.

Already, this method has shown that the rocks within the Earth are pierced by vertical columns where hotter and more fluid rock is rising upwards. When these rocks reach the bottom of the Earth's crust, they can spread out sideways, and form a soft layer of 'slushy' rock that the crustal plates can slide over. The top of a rising column of hot rock can also punch a hole through the Earth's crust, and belch forth as a volcano. The great volcanoes that stand isolated from the edges of plates are almost all hot-spot volcanoes: Hawaii, Tenerife, Saint Helena. As a plate moves over the hot-spot, the old volcanic cone is carried away, and the rocks push through to form a new erupting core. In the case of Hawaii, the old cones form a long sequence of islands and underwater seamounts stretching away to the north-west. The volcanoes of Venus and Mars seem to be the products of hot-spots in those planets; they have probably grown to such enormous sizes – relative to Hawaii, for example – because these two planets do not have moving plates, and all the volcano's rocky outpourings have built up on one spot.

The Earth's central core, revealed by the earthquake waves, is very dense, and it is almost certainly made of metal – probably the iron-nickel alloy of the kind that we find in some meteorites. Electric currents within the core generate a magnetic field that stretches way out into space all around our planet, enveloping it in an enormous invisible cocoon. This magnetosphere shelters Earth from the solar wind of charged particles that constantly flows from the Sun. When some of the particles do get in – especially at times of high activity on the Sun – the Earth's magnetic field guides them round to the regions of the poles, where they dive down and hit the top of the atmosphere.

From the orbiting Space Shuttle Challenger, astronauts photographed the volcanic island of Tenerife with a hand-held camera. The island and its volcano, Teide, arose over a 'hot spot' in the Earth's crust, where outflowing magma built up a conical land mass. Since then, Tenerife has drifted away from the hot spot. It will grow no larger, and Teide is now extinct.

The result is an aurora – a display of the Northern or Southern Lights. The energized atoms in the upper atmosphere glow as a rippling curtain of red and green light. The aurora is not only beautiful; it is reminder that our planet has a natural protection for life on its surface.

But it may be a good thing for some radiation to get through. Life on Earth has progressed only by the process of evolution, the gradual change in organisms that leads, just occasionally, to a more advanced kind of organism that does better than others in the struggle for survival. A small dose of radiation may be just what's needed to cause such a mutation.

Other cosmic influences have undoubtedly left their mark on the way that life has evolved on Earth. The intense cold of the Ice Ages is one example. There's no doubt now that the ebb and flow of the Ice Ages is caused by simple astronomical factors. A combination of the changing shape of the Earth's orbit, and the oscillating tilt of our planet's axis, means that the amount of contrast between summer and winter is continuously changing, over a period of thousands of years. When the seasons are mild, the summer sun is not strong enough to melt the ices at the Earth's pole, and the planet begins to head towards another Ice Age.

The most important event in the rise of the mammals on Earth – the death of the dinosaurs – may also have had a cosmic cause. Argument has raged about this for the past century: geologists contended it was a result of the Earth's changing climate, a cooling of the world that left the cold-blooded reptiles unable to cope. But in 1980 a team of American scientists, led by physicist Luis Alvarez, discovered something odd about the rocks laid down about 65 million years ago, at the time the dinosaurs died. They contain a thin layer of clay; and when Alvarez analysed the clay, he found it contained a lot of the chemical element iridium. This substance is quite rare on the Earth, but common in meteorites. The conclusion seemed quite natural. A huge meteorite had hit the Earth and exploded, distributing its material, rich in iridium, planetwide. The pall of dust raised shut out

Most nights, the polar snows are lit by the glows of aurora. This beautiful nineteenth-century painting shows curtains rippling above the north pole. The glows are caused when fast-moving particles from the Sun hit the top of Earth's atmosphere.

the Sun's light, so killing off the plants – and in turn, the great plant-eating dinosaurs.

Since 1980 the idea of a 'cosmic catastrophe' has gone further. Another pair of Americans, David Raup and John Sepkoski, looked at fossils of life in the sea, and concluded that our planet has suffered not just one great extinction, but several. Even more striking, the extinctions did not happen at random. They recur regularly, every twenty-six million years. An astronomical explanation would be the easiest way to explain the regularity. At first sight, however, there's a difficulty: twenty-six million years is much longer than the orbital time of anything in the Solar System, such as a swarm of meteorites that could regularly hit the Earth.

The answer must lie on the fringes of the Solar System. Here there is a ready-made barrage of missiles for throwing at the Earth: a swarm of icy snowballs, each waiting to become a comet when it approaches the Sun. If something perturbed these comet nuclei every twenty-six million years, a storm of them might fall to the centre of the Solar System, with many of them striking the Earth. Comet impacts would then be responsible for both the death of the dinosaurs (and the other extinctions) and the telltale layer of thin clay.

So, what is it that stirs up the comets in such a regular way? There are two popular answers. The more conventional explanation invokes our Sun's journey around the Milky Way Galaxy. As it goes round, with the Solar System in tow, the Sun bobs up and down, every 30 million years or so. And when it passes through the plane of our Galaxy, it's quite likely that the Sun will pass near enough to a dense cloud of gas for that cloud's gravity to perturb our comets, and so rain destruction on the Earth.

There could also be a more straightforward explanation. Suppose our Sun is not a single star, as we have always believed, but one of a pair of stars. The second star, Nemesis, is a long way off and is very dim – so astronomers have never spotted it. Now suppose that Nemesis orbits the Sun every twenty-six million years, following a path that brings it in just to the region of the comets. Then Nemesis would regularly stir up the comets, and fling them in towards the Earth.

So far, no one knows which of these theories is correct. Indeed, many geologists have disputed the evidence for regular extinctions of life on Earth; and some are proposing that even the evidence for a cosmic catastrophe killing the dinosaurs is far from conclusive.

But the whole controversy is a timely reminder that life on Earth is very vulnerable.

Above the Indian Ocean, Hurricane Kamysi wreaks havoc on the waters below. It is a reminder that we remain at the mercy of our environment, and that we should work with and not against it.

And now human beings have the power – wittingly or unwittingly – to affect the future of all forms of life on our planet. We are already doing so: as we have burned coal and oil, we have increased the amount of carbon dioxide in the atmosphere; at the same time, we are destroying vast areas of the tropical rain forests that could mop up this excess. America's Environmental Protection Agency has recently calculated that the build-up of carbon dioxide will have dire results. The gas will trap the Sun's heat, creating a greenhouse effect, and raise temperatures by as much as five degrees over the next 120 years. This would turn vast areas of fertile land into desert; and if the trend continues, we have only to look at Venus to see how our planet might end up.

Instead of this 'heat death', we could inflict a crippling cold on the Earth, if mankind ever resorts to nuclear war. American astronomer Carl Sagan has run a computer calculation of what would happen in a nuclear war, when the explosions would blow enormous amounts of dust up into the atmosphere. In broad terms, it's similar to the calculations on how an asteroid fall on the Earth could have killed the dinosaurs. But Sagan has gone into a great amount of detailed calculation, and he has checked his figures against the effects of dust raised by natural means – including the great eruption of the Mexican volcano El Chichon, and a dust storm that enveloped the whole of Mars in 1971.

Sagan's results are horrifying. Even quite a 'limited' nuclear war would cause a devastating 'nuclear winter'. Under the Earth's dust cover, the temperature would fall to minus 25°C, and stay that way for months. It could take years for the Earth to warm up again. Even without thinking of the effects of radiation, Sagan's calculations show that a nuclear war would have catastrophic consequences for life on Earth.

These lessons reinforce the fact that planet Earth is the only hospitable world in the Solar System. Our planet has evolved since its birth to suit living things; and life, in turn, has shaped the planet. The blue planet's 'oddities' – water, oxygen and life – are all interrelated. British scientist Jim Lovelock thinks that we should see the whole Earth as a living entity in this way: he calls the entity Gaia, after the Greek Earth goddess. Gaia seems to have had a 'sense of survival' that has lasted for billions of years. It must be in our interests now to work with her, not against her; and bear in mind that we have inherited that rarest of all things in the Solar System, a living world.

CHAPTER FIVE

THE MOON

21 June 1969 has gone down in history as one of those days when almost everyone can remember where they were and what they were doing. Who will ever forget those first blurred pictures of the Moon's surface live on television? That nailbiting wait as Neil Armstrong tentatively progressed, inch by inch, down *Eagle*'s ladder? The relief when he finally made it: 'That's one small step for a man, one giant leap for mankind.' Putting aside the politics and corner-cutting manoeuvres of the 'space race', the day, to most of us, was the start of a new era. It was the day we set foot on another world for the first time.

But there were others for whom an era ended. For many, the Moon was the epitome of remoteness, a romantic symbol on which to hang their dreams. The Moon of poets and playrights, the inconstant Moon shunned by lovers, the cold blue orb of artists – will never quite be the same again.

Strictly speaking, of course, the Moon is not a planet. It is a satellite of the Earth, like

Previous page: *Tiny and fragile, a manned lunar lander hangs in the sky above the inhospitable surface of the Moon. The series of manned Moon-landings captured the imagination of people the world over.*

To our ancestors, the Moon was an important goddess who lit the dark, dangerous nights, and they made up many myths about the markings they could see on its surface. This view of the Full Moon – seen from a rather unusual angle – was taken by the Apollo 11 astronauts as they headed back to Earth.

those which circle most of the other planets. But in our case – and to some extent in the case of Pluto and its moon, Charon – the situation is a little different. Both Earth and Pluto have only one, very large, moon: Charon is almost half the size of Pluto, and the Moon nearly a third the size of Earth. So the systems of Earth and Pluto can be considered more as a 'double planet' rather than just as a planet with a satellite.

To our remote ancestors, the Moon's appeal must have been anything but romantic. Their concern would have been far more practical. In the long, dark, dangerous nights, it was the Moon which lit the way. It must have been confusing and frightening, too, that the Moon's shape – and the amount of light it gave – changed from night to night. Sometimes it showed a thin crescent and was soon gone. At others it shone like a great silver ball the whole night long.

It's understandable that the early civilizations wanted to chart the Moon's changing phases. The Babylonians were the first to work them out in detail, and had such sophisticated knowledge that they were able to predict eclipses. The Greeks later applied that knowledge and improved upon it: they were able to explain *how* the phases were caused, and *why* eclipses should happen.

Like all the planets, the Moon has no light of its own – in fact, it is as dark as charcoal – and its phases are a direct result of its orbit about the Earth. The Moon completes one cycle of phases in 29½ days, beginning at new Moon. At this stage, the Moon is quite invisible. It lies directly between the Earth and the Sun, and so no sunlight falls on the side facing the Earth. A few days later, the Moon moves around in its orbit until a small sliver of sunlight illuminates its eastern limb (outer edge). The Moon then appears as a narrow crescent close to the Sun in the evening twilight.

As the month (a word which stems from 'moon') progresses, the Moon continues to move so that more and more of its surface is illuminated. Seen from the Earth, the Moon's crescent grows larger from night to night as it draws away from the setting Sun. Just after a week from new, we see a half-Moon in the sky. Because by now the Moon is a quarter of the way round its orbit, this phase is called 'first quarter'.

From now until day fourteen, the Moon's illuminated portion grows larger as it moves into a direct line with the Sun. And then, just over fourteen days from 'new', the Moon is 'full'. The Earth lies between the Sun and the Moon, and the Moon's face is fully illuminated.

After this, the waning Moon goes through its phases in reverse. Seven days after full there's 'last quarter', when the opposite half of the Moon shines in our skies, and then, once more in line with the Sun, the Moon is new again.

If the Sun and Moon were exactly in line *every* new Moon and full Moon, then we'd see eclipses of the Sun and Moon each month. As it is, the Moon's orbit is angled by 5°, and so exact line-ups are less frequent. Even so, there's always a handful of eclipses each year, although the line-up is so critical that you have to be in the right spot on Earth to see one.

Eclipses of the Sun are by far the most spectacular. They take place when the Moon is in its new position between the Sun and Earth. It is just a coincidence that the Moon and Sun, although very different in size, appear the same diameter in the sky. So the Moon can overlap the Sun completely, covering up its brilliant disc. When this happens, lucky people in a privileged band of locations across the world get treated to a total eclipse of the Sun. Because both the Earth and the Moon are moving, the duration of 'totality' is brief – never more than seven minutes in one place – but the memory lasts for ever. Day is plunged into night in seconds; the stars come out; birds and animals prepare to sleep; the temperature falls. But most spectacular of all is the sight of the pearly corona – the Sun's outer atmosphere – stretching out its ghostly fingers into the twilight sky.

Away from the exact path of totality (it's a 'path' because the Earth is spinning), an observer sees only a partial eclipse. Here the Moon overlaps just part of the Sun, and you can never see the corona – there is still far too much light.

Eclipses of the Moon are occasionally so subtle that you can overlook them altogether. They take place on the opposite side of the Moon's orbit, when it's behind the Earth as

seen from the Sun – and it moves into the Earth's shadow. When the full Moon enters the deepest part of the shadow, it sometimes disappears completely. But more often than not, it stays dimly visible as a coppery-red ball, even in the depths of eclipse. Sunlight is 'bent' around the edges of the Earth by its atmosphere, and this produces enough illumination on the Moon's surface to stop it from vanishing altogether.

But no matter how much the Moon's shape changed, there was always one respect in which it must, to our ancestors, have remained reassuringly familiar. It always keeps the same face towards the Earth – the familiar face of the 'Man in the Moon', seen by some as a hare, a kissing couple, or Cain with a bundle of sticks. Today, we know that this must be a result of the Earth's powerful gravitational pull, which has 'braked' the Moon's spin in the way that the Sun has nearly done to Mercury. In fact, the Moon does rotate on its axis once a month – if it didn't it wouldn't stay facing Earth as it circled us – and so all parts of its surface are exposed to the Sun. Even so, at fourteen Earth-days apiece, the Moon's days and nights are long.

When seen through a telescope, or even through binoculars, the familiar 'Man in the Moon' disappears in a welter of detail. Surprisingly, the first telescopic observer of the

By lucky coincidence, the Moon and Sun appear the same size in the sky, and so the Moon can overlap it exactly – causing a total solar eclipse. The path of the 7 March 1970 eclipse passed right over Wallops Island, Virginia, where an 85-ft (26m) spacecraft tracking dish is located. Above the dish, the eclipsed Sun's outer atmosphere shines brilliantly around the edge of the Moon. Galileo was the first to make proper records of what he saw on the Moon through a telescope. This 1610 drawing, made at first quarter, shows craters, mountains, and deep lunar basins.

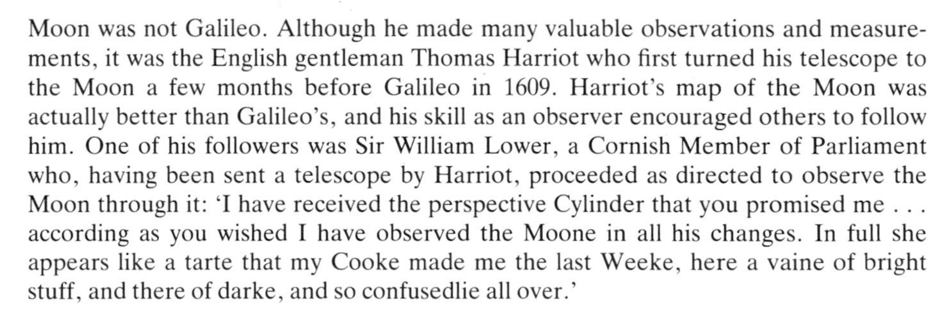

Moon was not Galileo. Although he made many valuable observations and measurements, it was the English gentleman Thomas Harriot who first turned his telescope to the Moon a few months before Galileo in 1609. Harriot's map of the Moon was actually better than Galileo's, and his skill as an observer encouraged others to follow him. One of his followers was Sir William Lower, a Cornish Member of Parliament who, having been sent a telescope by Harriot, proceeded as directed to observe the Moon through it: 'I have received the perspective Cylinder that you promised me . . . according as you wished I have observed the Moone in all his changes. In full she appears like a tarte that my Cooke made me the last Weeke, here a vaine of bright stuff, and there of darke, and so confusedlie all over.'

Galileo's accounts were altogether more scientific. He wrote: 'I have been led to the opinion which I have expressed, namely, that I feel sure the surface of the Moon is not perfectly smooth, free from inequalities and exactly spherical, as a large body of philosophers considers with regard to the Moon and other heavenly bodies, but that on the contrary, it is full of inequalities, uneven, full of hollows and protuberances, just like the surface of the Earth itself, which is varied everywhere by lofty mountains and deep valleys.'

Quite apart from upsetting the Church authorities in claiming that the supposedly perfect' Moon was 'full of inequalities', Galileo went on to show that, proportionally speaking, the Moon was an even more mountainous world than Earth. In typically ingenious fashion, Galileo realized that it was a simple problem in geometry to calculate the height of a mountain from the shadow it cast on the Moon's surface – all he needed to know was the angle at which the sunlight was striking. The heights of the mountain ranges on Earth and Moon are very similar, but then the Earth is over three times bigger.

Galileo's tiny telescope revealed the Moon's uplands, craters (which he described as 'spots') and low-lying basins (the Moon's 'darker part'). His conviction that the Moon was basically Earthlike was shared by observers who followed, who naturally concluded that the bright uplands were land, while the dark basins were seas. But what should the Moon's features be called? After a few unpopular attempts – one crater was briefly named 'The Greater Black Lake' – the Jesuit priest Giovanni Riccioli introduced the system we still follow today. His 1651 Moon map is a superb piece of draughtsmanship, with mountains, seas and craters all named. The mountain ranges were no problem. Riccioli named them after the Earth's ranges, and so we have the lunar Alps, Jura, Apennines, Altai and Caucasus. To the seas he gave romantic names: Mare Serenitatis, the Sea of Serenity; Mare Tranquillitatus, the Sea of Tranquillity; Oceanus Procellarum, the Ocean of Storms; and Sinus Iridum, the exquisitely pretty Bay of Rainbows.

When it came to the hundreds of craters which liberally peppered the Moon's surface, Riccioli was more controversial. He chose to name them after famous people (usually with astronomical connections) – a precedent which is followed even now in the namings of markings on planets and their moons. And so Plato, Hipparchus and Archimedes are three large and prominent lunar craters. However, Riccioli was by no means a supporter of the Copernican theory, and his hero Ptolemy got a huge crater right in the middle of the Moon's disc, while Galileo was only allotted a tiny pit on the far west of the Ocean of Storms – along with other infidels like Copernicus and Kepler!

Over the centuries, astronomers continued to probe the Moon with bigger and better telescopes. As finer details began to emerge, such as craters pitting the 'seas', it became obvious that the Moon was anything but Earthlike. The 'seas' were, in reality, dark, smooth plains, and completely dry. There were no clouds, no fog, no rain, in short, no atmosphere; the Moon's gravity is too weak to hang on to any. During the day the temperature of the unprotected surface rises to 105°C, while at night it plunges to minus 155°C – conditions far too extreme for life of any kind. The Moon appeared to be a lifeless world. On its surface, nothing ever changed.

It must have come as something of a shock to everyone in the mid-nineteenth century to hear – from no less an authority than Sir John Herschel, the distinguished son of Sir William (who discovered the planet Uranus) – that the Moon was actually teeming with life.

Between 1834 and 1838, John was busy working in South Africa near the Cape of Good Hope. There he had taken a large eighteen-inch diameter reflecting telescope, made by his father, in order to carry out the first ever detailed survey of stars and other remote objects visible in the skies of the southern hemisphere.

The result of an impact by a giant meteorite, the crater Daedalus measures some 50 miles (80km) across. The crater, on the Moon's far side, was photographed from lunar orbit by the astronauts on Apollo 11. Other craters are hundreds of miles across, while some are no more than pits.

Although John's thoughts could hardly be further from the Moon, a cunning newshound on the *New York Sun*, Richard Locke, decided that the survey had the makings of a good story. Furthermore, communications weren't too reliable in those days; and Locke surmised that Herschel wouldn't be able to get back at him until the rumpus he was about to create had died down.

The astonished readers of the *Sun* picked up their newspapers on 25 August 1835 to be confronted with the headline: 'Great Astronomical Discoveries made by Sir John Herschel at the Cape of Good Hope.' There followed an account of how Herschel had built an enormous telescope which revealed details on the Moon never seen before. Over the following week, Locke went on to tell the *Sun*'s readers about the amazing discoveries Herschel had made. Quite apart from the incredible scenery of 'oblisk-shaped or very slender pyramids . . . of a faint violet hue, and very resplendent', the Moon was inhabited! The 'intelligent life' appeared to consist of bat-men 'four feet in height . . . covered with short and glossy copper-coloured hair . . . faces of a yellowish flesh-colour, a slight improvement upon that of the large orang-outang'. But there were other lifeforms too – notably 'a strange amphibious creature of spherical form, which rolled with great velocity across the pebbly beach.'

Public reaction must have been akin to that stirred up by the first manned lunar landing. The *New Yorker* claimed that Herschel's work had 'created a new era in astronomy and science generally', while even the more cautious *New York Times* maintained that the discoveries were 'both probable and plausible'. But the hoax was soon exposed for what it was by a rival paper and Locke himself was forced to come clean on 16 September. Poor Herschel wasn't himself able to issue a denial until weeks later, when the news finally reached South Africa. However, the public reaction and the *Sun*'s soaring circulation proved that people were as desperate then as they are now to believe that there must be life of some kind 'out there'.

The Moon hoax was by no means the first – or last – crackpot story pinned on our unfortunate satellite. There's still a Russian theory that the Moon is really a hollow spaceship, placed there by aliens; an American claim that its surface is under excavation by an intelligent race; and another American assertion that the astronauts 'knew' about the aliens, and that their space-jargon to Mission Control was actually coded information about 'them' . . .

By the beginning of this century, astronomers were agreed about most aspects of the Moon. Hoaxes aside, they had concluded that it was a small, dead, airless world, subject to unrelenting temperature extremes. But its surface was pitted with craters of all sizes, ranging from the huge circular maria basins hundreds of miles across down to tiny ones just visible in the world's biggest telescopes. Where had they all come from? Astronomers concluded that they were either dead volcanoes or the result of a tremendous bombardment by meteorites. On balance, although both processes must have contributed to the lunar landscape, it seems that bombardment has been by far the more important.

This century, the Moon has been used for a different kind of target-practice. At a distance of a mere 238 000 miles (380 000km), it was a natural goal to be reached after we had proved we could launch probes and satellites into space. After proving they could hit the Moon with Luna 2 in September 1959, Soviet space scientists followed just a month later with Luna 3, which sent back the first photographs of the Moon's farside. Bearing out predictions, it turned out to be very heavily cratered, but with few 'seas'. Several of the craters have since been named in honour of Russian scientists. But long before any probes flew behind the Moon, the mountainous ridges of the biggest of the basins on the farside had already been mapped by observers from Earth, who saw them in profile at the Moon's eastern limb (edge). Recognizing that they must be the walls of an enormous 'sea', moon-mapper and discoverer Patrick Moore christened the feature Mare Orientale – the Eastern Sea. (As he wryly observes, however, it is now on the western side of the Moon! For a long time, astronomers had mapping all their own way, which meant that co-ordinate systems for lunar and planetary maps were upside-down and back-to-front to correspond with the view you

get in a telescope. When probes and people reached the Moon – the right way up, of course – it all had to change.) But Eastern Sea or Western Sea, the Mare Orientale is a spectacular bullseye, bearing a very striking resemblance to the Caloris Basin on Mercury.

The Russians sent two remote-controlled Lunokhod rovers to the Moon. Each 8-wheeled vehicle spent several months on the lunar surface, photographing the landscape and analysing a variety of rocks.

In the 1960s, the accelerating build-up in the manned moon-race led to many spin-offs, in the shape of detailed maps, close-ups of its surface and surveys of the lunar environment. While the Soviets concentrated on an attempt to softland a craft on the Moon (which they achieved with Luna 9 in 1966), the Americans sent a series of kamikaze-style Ranger probes which returned a sequence of pictures before crashing into the surface. These they followed with soft-landing Surveyors, which analysed soil samples and returned surface photographs. In the late 1960s, both space superpowers made detailed photographic surveys of the surface from Moon orbit, while they continued in parallel with their manned programmes. American astronauts and Soviet cosmonauts alike were sent to 'moonlike' places on the Earth to train in readiness for the real thing.

Once the Americans got there, the race was over. There was only one decent thing the Russians could do, and that was to pretend they had never intended to send a man to the Moon in the first place. It was an unnecessary risk of life, they argued, and a very expensive way to collect rock samples. To prove their point, they went on to send a number of unmanned Luna probes to the Moon which automatically returned small rock samples to Earth for analysis. Two of these probes carried a remote-controlled moon-car, the Lunokhod, which travelled the surface transmitting TV pictures of its surroundings. Lunokhod 2 (which landed in January 1973, after the Apollo missions were over) covered 23 miles (37km) of the lunar surface, and was able to analyse in detail many of the different rocks it encountered.

But it was the astronauts who captured the world's imagination. All six manned lunar landings took place in a period of less than 3½ years, from July 1969 to December 1972. Seven missions were, in fact, planned, but Apollo 13 had to be aborted. Two days into the mission, an oxygen tank aboard the spacecraft exploded, leading to an immediate loss of electrical power. Desperately low on oxygen and with only meagre power supplies left, the three astronauts were still able to manoeuvre their craft behind the Moon – it was too late to turn back – and return it safely to Earth. As an ironic bonus, they were able to get some of the best photographs of the lunar far side ever taken.

Otherwise, the Apollo missions each followed the same pattern. Blasted off in their craft atop a huge Saturn V rocket, the three astronauts made for lunar orbit. While one astronaut stayed aboard the orbiting command module, the two others would crawl into the lunar module which later separated and then landed at the appointed spot on the Moon. After a period spent collecting rock samples and laying out experiment packages on the surface, the two astronauts would blast off from the Moon to dock once again with their orbiting colleague in the command module and return to Earth.

As it grew more confident, America's NASA Space Agency increased the sophistication and duration of the Apollo missions. Neil Armstrong and Buzz Aldrin, the first two astronauts, were on the surface for under a day and collected 46lb (20.8kg) of rock samples. Apollo 17's Gene Cernan and Jack Schmitt – the latter a qualified geologist – stayed for 12 days 14 hours, and brought back a record load of rocks weighing 257lb (116.5kg). The last three Apollo missions made extensive use of the Lunar Rover, an electrically powered car which travelled at speeds of up to 8mph (13kph) over the Moon's uneven surface. In it, the astronauts were able to see much more of the Moon and sample the soil from a far wider range of environments.

Apollo 17's Harrison 'Jack' Schmitt – the only geologist to have visited the Moon – examines an enormous boulder near the landing-site in the Taurus-Littrow Mountains. The Lunar Rover, in which he travelled the Moon's surface, is to the right of the boulder.

From studies of moonrock picked up from a wide range of sites, and from records of seismic equipment left on the surface to monitor 'moonquakes', it's been possible to reconstruct something of the Moon's past life. The rocks tell us that the Moon was born at the same time as the other planets, 4.6 billion years ago. Chunks of rock pounded into the young Moon, slowly building it up to its present size and melting the surface layers with the heat of their impacts. Light rocks floated to the top of this molten surface as a kind of scum. As the bombardment from space slackened off, the scum solidified into the highlands we see on the Moon today. The remaining meteorites – too few in number to melt the surface wholesale – simply left colossal scars where they landed. At or around this time, the really big meteorites – perhaps the Moon's own moons – also hit the surface and excavated the huge maria basins. They were filled by molten lava welling up from deep inside, which cooled and solidified into the dark basaltic plains we see today. Since those days, over three billion years ago, very little appears to have happened. The reservoirs which flooded the basins have now solidified, and the only hot, liquid parts of our Moon are deep within its core. The shifting, molten rock *does* give rise to moonquakes – but they are puny compared to those on Earth. The total amount of energy released by all the moonquakes in one year is less than that involved in a good-sized firework display!

In short, the evidence from Apollo tells us that the Moon is dead. It is a fossil, its surface unaltered and preserved for three billion years. On planets like Earth, Mars and Venus, the ancient record has been destroyed by erosion, volcanic activity and – in the case of the Earth, at least – continental drift. The Moon is a repository of the ancient history of the Solar System.

As well as telling of the Moon's past, the Apollo experiments can tell us a little about its future. One experiment has shown quite conclusively that the Moon is moving away from us. Since the Apollo landings in 1969–72, the Moon is about 1½ feet (½ metre) more distant.

The reason lies in the fact that the Earth and Moon are close enough to raise tides upon each other. The two tidal 'bulges' the Moon raises in the Earth's oceans act like a pair of brake shoes, and between them our spinning Earth is gradually slowing down. The increase in the length of the day hardly seems significant – 1 second in every 50 000 years – but it can be measured with modern atomic clocks.

At the same time, the water in the tidal bulges has a tiny gravitational effect on the

Astronaut Jack Schmitt stands by the US flag while, high above, the Earth shines brilliantly in the dark lunar sky. Photographer Eugene Cernan can be seen reflected in Schmitt's visor.

Moon. They act so as to pull the Moon ahead in its orbit – which means that the orbit grows slightly bigger. The result is that the Moon moves further away.

Although this effect was known before the days of Apollo, we now have a much more accurate measurement of it, thanks to the 'corner reflectors' the astronauts left behind on the lunar surface. By beaming a powerful laser pulse from Earth off a reflector and timing its return – a matter of only 2½ seconds – it's a relatively simple matter to measure the Moon's distance to the accuracy of just a handsbreadth (4 inches or 10cm). These laser ranging experiments have shown that the Moon is definitely moving away.

Where will it all end? If you work through the calculations, it turns out that billions of years from now, our day will be fifty-five of our present days long, as will be the length of the month. The distant Moon will hover over one spot on the Earth's surface, so that our distant descendants on the opposite side of the globe might never see it. They, and people living directly under the Moon, would have permanent high tides.

Although we can predict how the Moon might end, and how it was born, what the Apollo missions do not tell us is where the Moon came from. Analysis of the lunar rocks reveals puzzling differences between those of Earth and Moon. In particular, the rocks are comparatively rich in the elements titanium, uranium, calcium and iron; but quantities of gold, silver, zinc and potassium are low. If the Earth and Moon were formed together in space, then the rocks should be more similar.

One possible explanation is that the Moon was born in a different part of the Solar System from the Earth, and that the Earth's gravity later 'captured' it. This would explain the different rock composition, but many astronomers are unconvinced by such a theory. Another idea – first put forward by George Darwin (second son of Charles) in 1879 – suggests that the Moon was a piece of the young Earth which broke away. Supporters of this idea claim that the moonrock has a composition similar to earth rock deep-down. Once again, a number of astronomers are uncomfortable about the idea that parts of the Earth might 'fission off'. Then there's the suggestion that the Moon and Earth *were* born together in space – but if so, why don't other planets (with the exception of Pluto) have single, Moon-sized satellites?

To answer these questions, scientists plan to go back to the Moon. 'I believe it likely that before the first decade of the next century is out, we will, indeed, return to the Moon,' declared NASA's Administrator James Beggs at a conference in 1984. His optimism was echoed by Presidential science advisor George Keyworth who maintained that a lunar base was 'one of the more obvious of the bold and exciting goals that we can reach' in the near future. NASA is looking, too, for international co-operation to help fund the venture. Many of the 300 delegates see the development of moonbases as resembling the growth of research programmes in Antarctica, where there is a rule of peaceful international co-operation.

This sixteenth-century German engraving represents the Moon. The scene shows a tidemill, near which fishermen cast their nets. The link between the Moon and the tides has been appreciated throughout history.

Quite apart from the research potential – a magnet to astronomers and geologists alike – the first moonbase would justify its expense by industrial operations. The high setting-up cost would mean that it wouldn't return a profit for many years. But the low lunar gravity would allow cheap shipment of metals to Earth orbit, where they could be used for construction of items on a large scale; and those metals which are common in moonrock, such as manganese, aluminium and magnesium, could be smelted in commercial quantities.

'The next time we go to the Moon, it will be to stay,' declared one delegate fresh from the conference. So when will it be? Optimists see no reason why a base shouldn't be established by 2001. In the meantime, there is enough to be getting on with, for there are still questions to answer. For example: is the Moon *completely* dead? Keen amateur astronomers who have the dedication and the leisure to watch the Moon every night sometimes report localized transient glows; what are they? Could they be releases of gas? Dust stirred up by landslips? Volcanic activity? It's true that no photograph of a 'transient' has ever been secured – but many professionals think the amateurs may have a point.

So it could be that we don't know the Moon quite as well as we once thought. We still don't know where it came from. We don't know why Earth and Moon are virtually unique in forming a double-planet. We can't even be sure that the Moon is utterly dead. So let us echo George Keyworth in support of a moonbase. 'The space program,' he maintains, 'ought to be both practical *and* visionary.'

Plans – on paper at least – are underway in the US for a Moonbase, to be constructed in the very near future. The base, scheduled for operation by the beginning of the 21st century, would use the low gravity and the vacuum of the Moon for ambitious industrial projects. This artist's impression shows a recently-landed Moonbase module about to be ferried off to join others at the growing base.

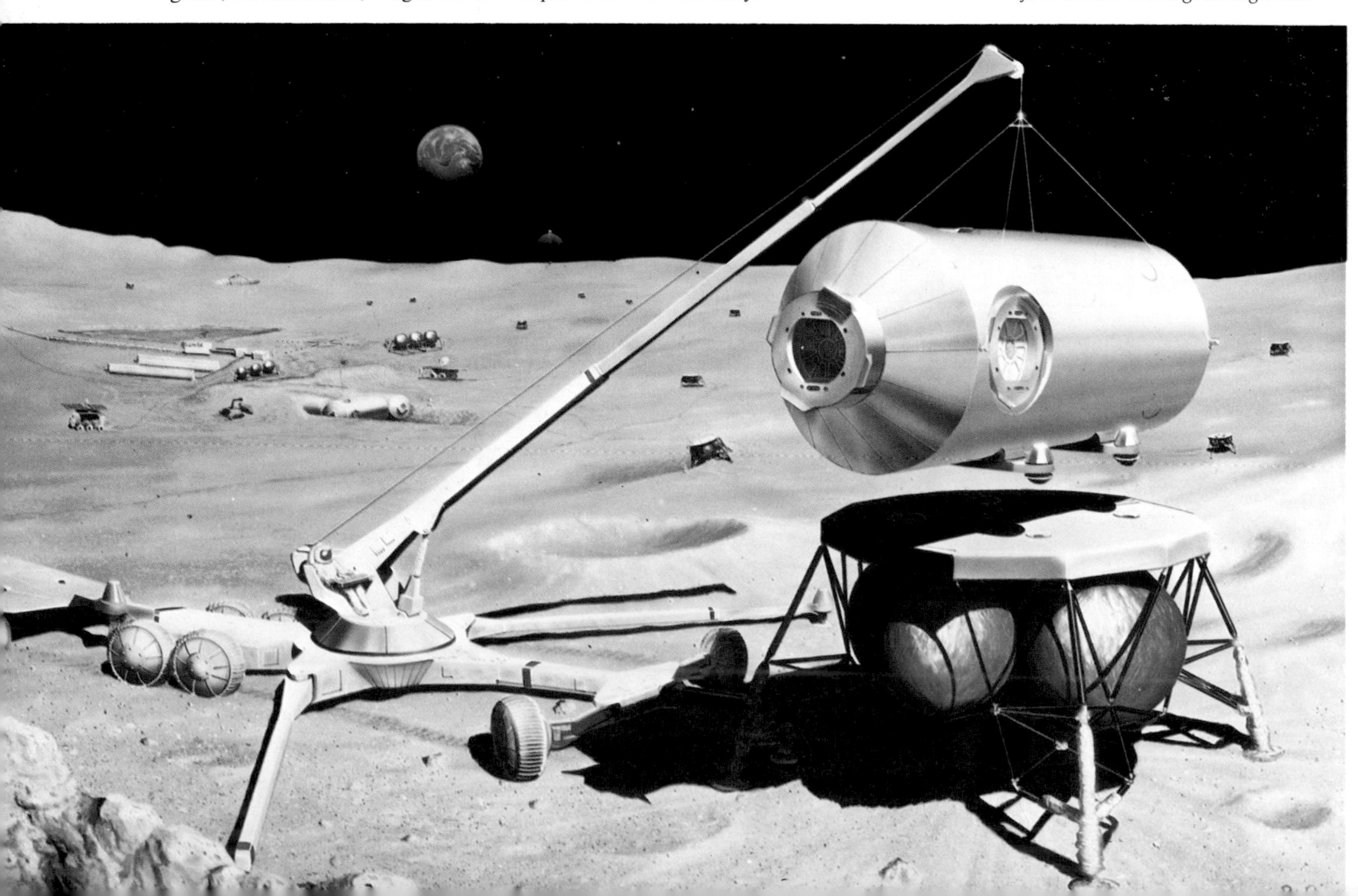

CHAPTER SIX

MARS

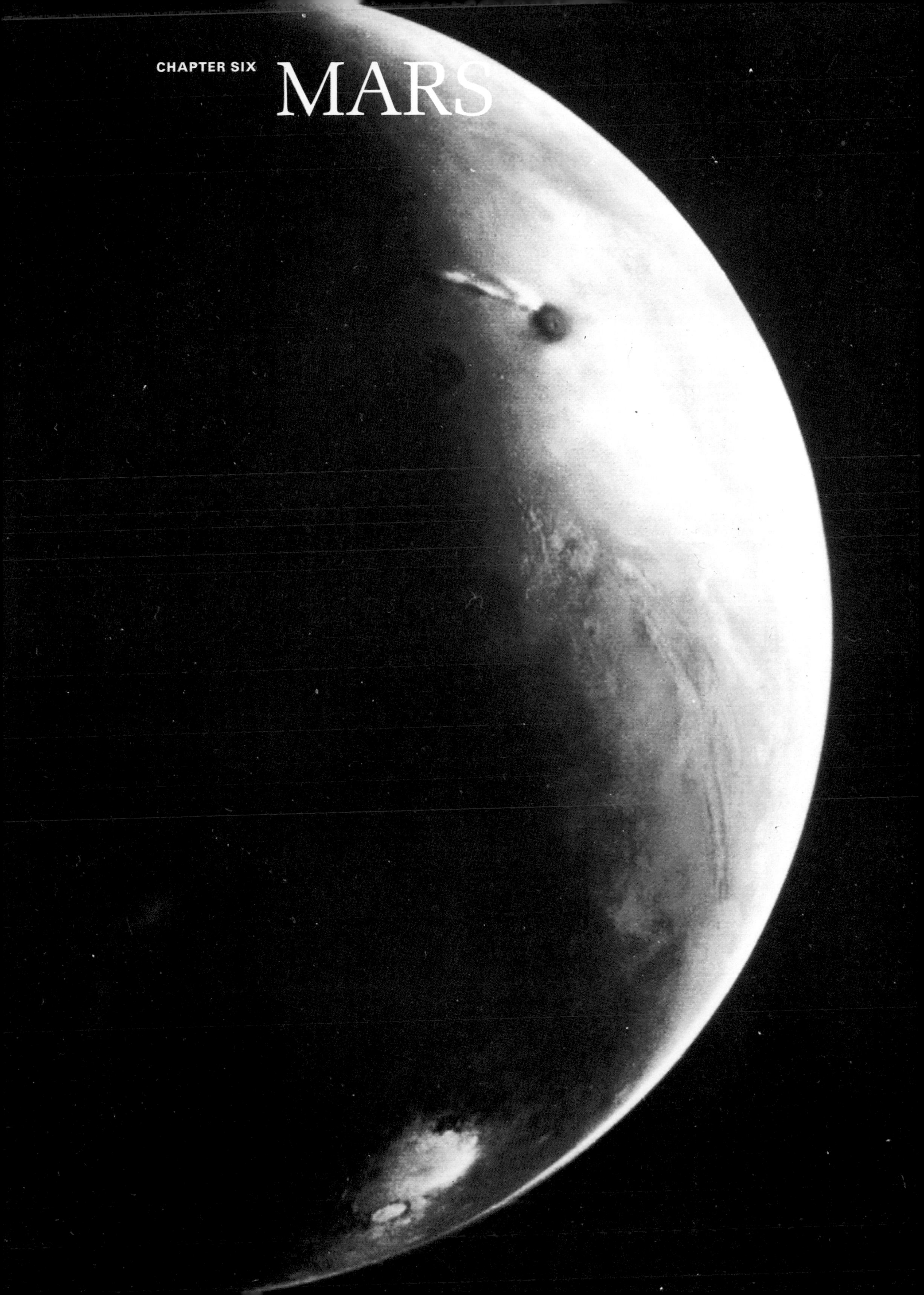

Martian terror strikes the Earth, in the 1953 film of H. G. Wells's The War of the Worlds.

'Ladies and gentlemen,' began the radio presenter, solemnly. 'I have a grave announcement to make.' People who heard these words, almost half a century ago, have never forgotten them. Coming just after eight o'clock in the evening on Sunday 30 October 1938, they sparked off one of the biggest scares the world has ever known. For one night, nearly half the population of North America believed it was about to be destroyed – by Martians.

The whole affair was actually a rather-too-realistic stunt to improve audience ratings for CBS radio. It was masterminded by the brilliant actor, director, and one-time bullfighter and magician, Orson Welles – then aged twenty-three – who had the responsibility for putting on a series of Sunday evening radio plays. His problem was that people much preferred to tune their wireless sets to the Charlie McCarthy Show on one of the rival networks instead. Welles finally realized that he'd have to put on something spectacular to compete – or his days in radio, along with those of his Mercury Theatre troupe, would be numbered.

Welles had two big points in his favour that night. One was his choice of play: a dramatized version of H. G. Wells's *The War of the Worlds*. The other was the fact that Charlie McCarthy had an unknown singer on his show, prompting bored listeners to start twiddling their dials. So a much bigger audience than usual heard the announcement: 'The strange object which fell at Grovers Mill, New Jersey, earlier this evening was not a meteorite. Incredible as it seems, it contained strange beings who are believed to be the vanguard of an army from the planet Mars.'

The announcement was followed by soft music, which served only to heighten the tension. Then the announcer broke in again, his voice showing the faintest trace of panic. The Martians were spreading out, he reported. They were horrible, scrawny, leather-skinned creatures. More music, more urgent announcements followed. There were chilling silences as the announcer waited for information. A voice from Washington then revealed that the Martians were landing all over the United States. Thousands of people had already been slaughtered in cold blood by the aliens'

Previous page: *The Viking 2 Orbiter show Mars as a crescent, with the ice of the south polar cap (bottom) and part of the system of canyons that girdles Mars's equator. To the north is the volcanic Tharsis plateau, where a plume of icy cloud is blowing off the peak of the volcano Ascraeus Mons.*

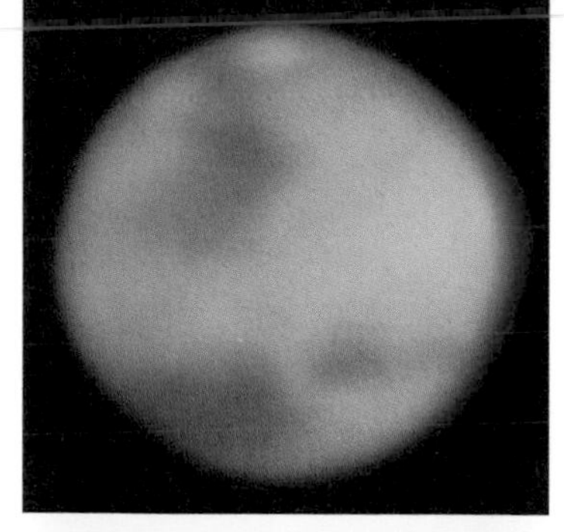

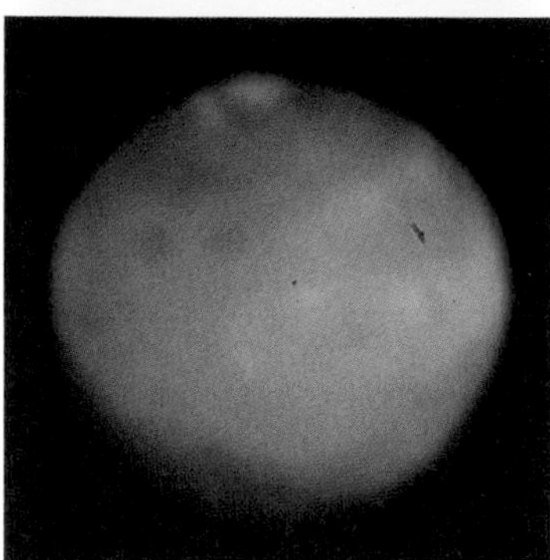

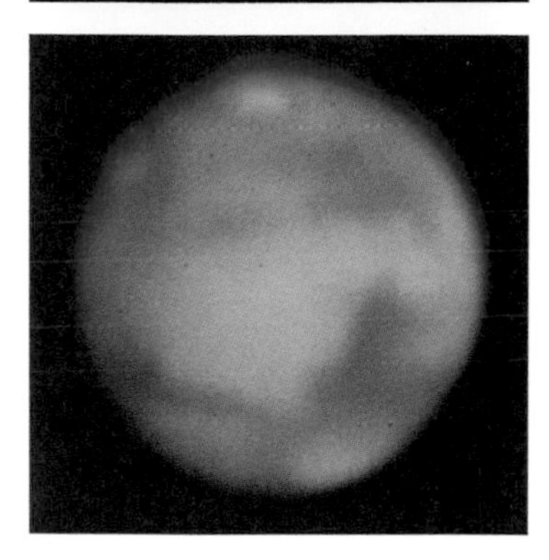

Photographs of Mars taken from the Earth reveal greyish-green markings on red deserts, with a white polar cap. This sequence of pictures shows the planet rotating, once in just over 24 hours. These characteristics make it the most Earth-like of the other planets.

death-ray guns. Finally, listeners heard the announcer's panic-stricken cries as the Martians swept through Manhattan and into the radio station itself. The broadcast ended with a chilling, high-pitched scream.

People who stayed with the play to the end found out that they were listening to fiction. But thousands had left their wirelesses before the play finished, in a state of blind panic. Some were making for the hills with what they could snatch. Others wrapped wet towels around their heads to absorb the noxious gases. People prayed in the streets; bars and restaurants emptied; and a few people even claimed to have *seen* Martians.

Orson Welles knew nothing of the panic he had caused until the next morning, when he saw his name in lights outside the *New York Times* building. 'Orson Welles Causes Panic' read the neon news display. Welles bought a newspaper, only to be confronted with headline: 'Radio Listeners in Panic: Many Flee Homes to Escape Gas Raid from Mars'. As America settled down to rest its jangled nerves, Welles received a severe ticking-off from the newspapers for the realism of his presentation. People even brought lawsuits against CBS totalling $750,000; but eventually all were withdrawn. Welles's career, however, was assured. CBS was absolutely delighted to have America's most notorious actor broadcasting on its airwaves, and Welles's contract was never in danger again.

Belief in Martians dies hard. Of all the planets, Mars is the one most popularly associated with having life of some description on its surface. Who hasn't heard of the Martian canals? And well into the 1960s, a number of astronomy books were confidently predicting that probes would find some kind of primitive plant life on the planet.

Mars shines blood-red in the skies of Earth, and is aptly named after the god of war. Although it's the second closest planet – it can approach to a distance of about thirty-four million miles (55 million km) – it ranks third in brightness after Venus and Jupiter, because of its small size. At just over half the size of Earth, Mars can never look as awe-inspiring as Venus does among the stars. But it makes up for its comparative dimness – it is still generally brighter than most of the stars – by its striking colour. Added to this there's the fact that Mars, being outside the Earth's orbit, is not constrained to follow the Sun closely, and can be seen high in the sky at the dead of night.

Mars, viewed through a small telescope, is a great disappointment. All you can see is a tiny red disc. It needs patience – and a telescope with a mirror more than six inches (15cm) across – to make out the details. But astronomers who were able to observe the planet began to build up a picture of it that was remarkably Earth-like.

First, there was the length of Mars's 'day'. As long ago as 1659, the Dutch observer Christiaan Huygens – the first person to realize the true nature of Saturn's rings – noticed that a V-shaped marking appeared regularly on Mars's disc. From this, he concluded that Mars rotated roughly once every twenty-four hours. The exact value is now known to be 24 hours 37½ minutes.

Then there was the tilt of Mars's axis. The Earth's axis is tilted at 23½° to its axis of rotation. William Herschel discovered that at 23.98°, Mars's tilt is almost identical. A planet's tilt determines its seasons – it all depends which hemisphere is pointing towards the Sun – and so there was every reason to expect the seasons of Mars and Earth to be similar, too. The difference would be that with a year lasting for nearly two Earth-years, each Martian season would be twice as long as ours.

Quite apart from the similarities in Mars's day and in its seasons, its very appearance was Earth-like. At its poles were white markings. Over the course of the Martian year, these were seen to grow and shrink in exactly the same way as the icecaps on Earth. It was quite natural to assume, as William Herschel did, that these were indeed polar caps made of ice and snow. On Mars's disc itself, bigger and better telescopes were able to pick out a complicated pattern of dark markings. At first it was thought that the dark areas were seas and the red plains dry land, as had been thought in the case of the Moon. But – as observations by Herschel and others found – Mars's atmosphere was far too thin to sustain seas. Furthermore, the seas were moving! Over the Martian year, some of the markings apparently darkened and grew larger as the 'summer' approached, only to 'die back' to their former extent in the 'winter'. It was very natural to conclude, as the French astronomer Liais did, that the markings were old seabeds now filled with vegetation – primitive seaweed, perhaps.

The only problem with this Earth-in-miniature was that astronomers got few chances to see it in close-up. Mars has a very elliptical orbit, which means that even when it and the Earth are side-by-side in their paths around the Sun – at opposition – the separation can be as much as sixty-five million miles (102 million km). But at the best possible oppositions, when Mars is at its closest to the Sun and the Earth is furthest away, the separation can shrink to only thirty-four million miles (55 million km). In 1877, astronomers all round the world readied themselves for one such close encounter.

One of the first surprises was the discovery of Mars's two tiny moons. They were found after a systematic search by Asaph Hall, using a new telescope at the US Naval Observatory in Washington. In fact, Hall very nearly abandoned his hunt when the glare from Mars's surface proved too great. He put the telescope to bed, packed up and went home. But his wife insisted he gave it one more try. The very next night, Hall picked up a tiny, moving object next to the planet – and then the clouds moved in. It was five frustrating days before Hall had the chance to check. To his (and his wife's!) delight, not only *was* the object a moon but the very next night, he found another one, too. The moons, both little potato-shaped lumps no more than fourteen miles (22km) across, were named Phobos (fear) and Deimos (panic) after the attendants of the war god. In honour of the important role Hall's wife had in their discovery, the biggest crater on the larger moon (Phobos) has now been given her maiden name – Stickney.

Boston businessman Percival Lowell built this fine 24-inch (60cm) telescope at his observatory near Flagstaff, Arizona, specifically to observe Mars.

One observer making a special study of Mars at that time was the Italian astronomer Giovanni Schiaparelli. From his observatory in Milan, he released new maps of the planet which showed details never glimpsed before. In particular, Schiaparelli's maps revealed that the dark markings on Mars were connected up with an extensive network of long, fine, straight lines – or *canali*, as he called them. In Italian, *canali* means 'channels'; but the word rapidly became corrupted to 'canals', with all their artificial implications. At the next opposition in 1879, Schiaparelli again saw the *canali*, and this time recorded that a few of them appeared double for short periods – hardly a 'natural' kind of thing to happen. Other observers didn't start drawing canals until about six years later, but after that many more astronomers joined in. From then on, the existence of the canal network was never in any doubt in some quarters, and nor was the interpretation put on it. It was clearly a last-gasp attempt to save a dying planet. Mars was desperately short of water – and so the brilliant Martian engineers had constructed a global canal system to convey water from the seasonally-melting polar caps to the deserts around the equator. The Martians themselves lived in cities located at 'oases', where two or more canals met.

The majority of astronomers couldn't see any canals at all, and disagreed entirely with this romantic interpretation, yet it had tremendous popular appeal. It had a powerful and charismatic champion in Percival Lowell. Lowell was a member of the prestigious Boston Lowells – a family so distinguished that it was said 'The Lowells speak only to Cabots, and the Cabots speak only to God.' Like all the Lowells, he was refined, cultured and superbly well-educated. In the late 1880s, when he learned that Schiaparelli's sight was failing, Lowell determined to devote himself to the study of the Martian canal system. With the help of his friend Andrew Douglass (who later became famous for establishing the ring-counting system for dating trees), he set up an observatory dedicated to the study of Mars on a 7500 foot (2210 metre) pine-covered plateau at Flagstaff, Arizona.

Lowell made countless drawings and photographs of the Martian surface. His beautiful series of Martian globes, housed in what is now the Lowell Observatory's main hall, show how much firmer his conviction about the canals grew with time. The later maps and globes show over 200 canals, all linear and very artificial in appearance.

As his conviction grew, so too did his desire to tell the world about the plight of the ingenious Martians. It was evident to Lowell that the Martians must be globally united and free from the scourge of war – something which he, as a pacifist, was very keen to broadcast to the people. But the more he said, the more he alienated the Lowell Observatory from the mainstream of astronomy.

Lowell saw hundreds of canals on Mars – thin straight lines that joined the larger dark areas. In November 1894, he depicted these two canals (Phison and Euphrates) as double, with a large 'oasis' where they cross. Lowell was convinced the canals were the work of intelligent Martians; in fact, they were an optical illusion.

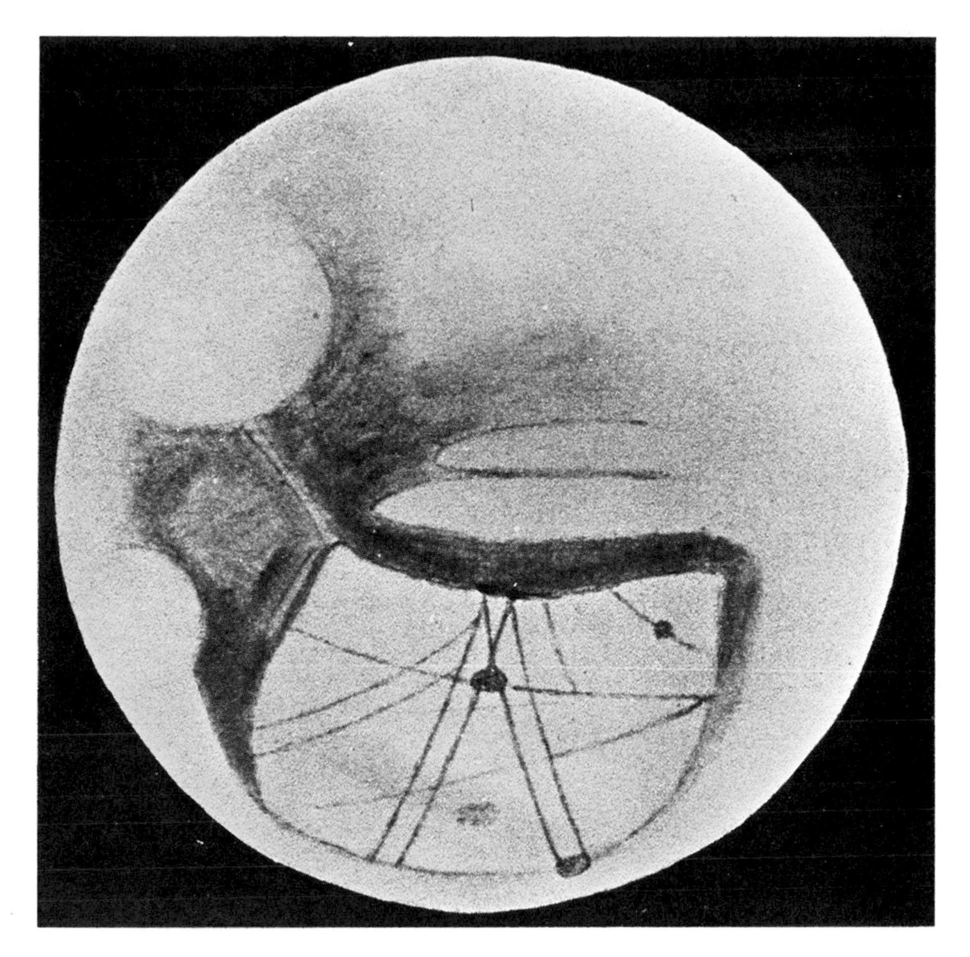

Why did Lowell see canals, while most other astronomers didn't? Clyde Tombaugh, who was later to discover the planet Pluto, believes that it may have been an effect caused by Lowell's telescope, which he himself has subsequently seen. According to his observation notes, Lowell felt he was getting the sharpest images of Mars (without the haze of rainbow colours that all lens telescopes produce) when he cut down the telescope's lens with a diaphragm to only sixteen inches (41cm). 'Stopped down' like this, the telescope gives very bright images. If there are two bright patches on the planet separated by a dark marking, the effect will be for the bright patches to 'bleed' into the dark, making the dark marking a lot narrower. Clyde Tombaugh has managed to see some quite convincing 'canal' networks in this way – but they are completely illusory.

Nevertheless, belief in Martians dies hard, as the *War of the Worlds* scare showed. And as we have seen, there were still a number of astronomers in the early 1960s who held out hope that Mars would have some primitive form of life.

So many observers all round the world were bitterly disappointed with the results of the first spaceprobe mission to Mars in July 1965. The US probe Mariner 4 flew past the planet at a distance of 6120 miles (9800km), and photographed – craters. Making allowances for the slight erosive effects of the thin Martian atmosphere, Mars's surface appeared to be almost identical to that of the Moon. Not only that, but there was far less by way of atmosphere than had been thought; the atmospheric pressure was less than one-hundredth that of the Earth, and the main constituent was unbreathable carbon dioxide. Even more disappointing were the dark markings. These were simply regions of lower surface brightness, periodically covered and uncovered by the seasonally-shifting windblown Martian soil. What had seemed from Earth to be patches of vegetation growing and shrinking turned out to be areas prone to invasion by sand.

Mariners 6 and 7, which in 1969 flew by Mars at a distance of a mere 2000 miles (3500km), only confirmed the barren picture. Their photographs once more showed a dead, cratered landscape. But the Americans tried again, and in 1971 they sent Mariner 9 into orbit about Mars – the first probe to go into orbit around another planet. The idea was that Mariner 9 should conduct a detailed photographic recon-

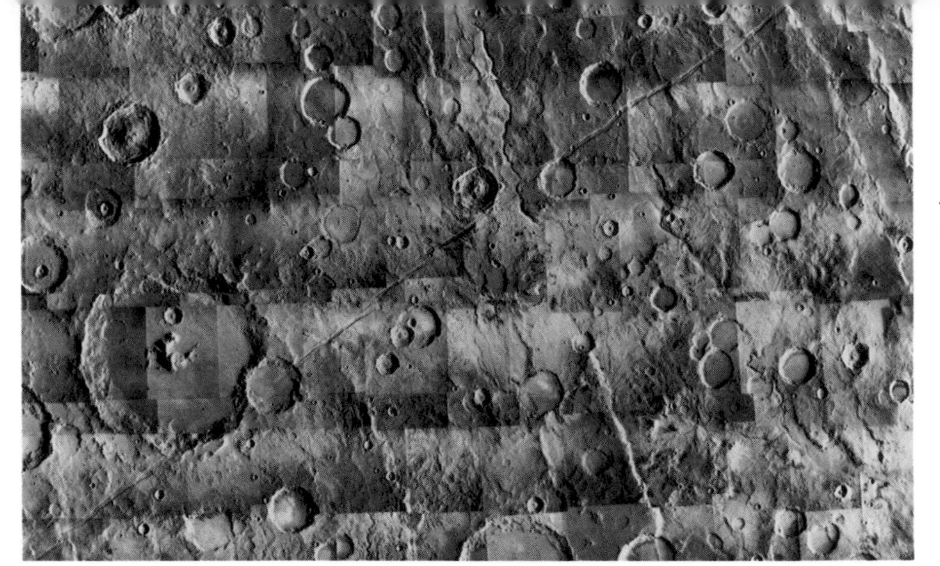

Half of Mars is covered with craters, and looks very like the Moon's surface. By ill-chance, the first probes to fly past Mars all photographed terrain like this, giving the impression that Mars was not only lifeless, but a totally uninteresting world.

naissance of Mars. But for several weeks, there was nothing to be seen. The whole planet was wrapped in a global duststorm.

As the dust began to settle, the waiting scientists were astonished to see what looked like the summit of a fifteen-mile (24km) high mountain peeking above the clouds. With the continuing clearance, they were even more amazed to discover that their mountain was in fact a huge 'shield' volcano, measuring 350 miles (560km) across. Soon a whole plateau of volcanoes was unveiled for all to see. As the clearing continued, Mariner 9's cameras started to pick out great cracks in the Martian landscape. One system of chasms, named the 'Valles Marineris' in honour of the probe, made a 2500 mile (4000km) long gash across the planet's equator – ten times longer, six times wider and four times deeper than the Grand Canyon!

But of all the unexpected sights revealed by Mariner 9, it was the 'sinuous channels' which took pride of place. These narrow twisting valleys – orders of magnitude smaller than the 'canals' – are incised into the rocks in such a way that they can have been cut by only one medium: running water. Now they are completely dry; but they are evidence that water once flowed there. But if there ever *was* water, then life – which relies on water for the chemical reactions responsible for initiating and sustaining it – may once have flourished on Mars.

Suddenly the scientists were incredibly optimistic again. Mars wasn't dead; far from it. It was just one of those unfortunate things that the first three fly-by Mariners had been targeted past the most cratered and unexciting part of the planet. Buoyed up with enthusiasm and optimism, astronomers, chemists, biologists and engineers alike joined forces to build the two probes which would – they all hoped – end a century of speculation for good.

Seven years to the day after Neil Armstrong and Buzz Aldrin set foot on the Moon, a small but highly-sophisticated unmanned spacecraft touched down safely on the salmon-pink plains of Mars.

The Viking project was one of the most complicated missions ever mounted. It involved four spacecraft – two orbiters and two landers. Two orbiter-lander combinations were to travel to Mars, arriving within months of one another. Once in Mars orbit, the combinations would separate. While the orbiters continued to make a detailed photographic reconnaissance of the Red Planet, the landers would descend – slowed first by parachute and then by retro-rockets – towards the Martian surface.

Both Viking missions were launched from Cape Canaveral by Titan-Centaur rockets in the summer of 1975. Viking 1 blasted off on 20 August followed on 9 September by Viking 2. In anticipation of an encounter with some kind of extraterrestrial life, both landers had been sterilized prior to launch to ensure that they would cause no contamination – either to the Martian 'bugs' or to the Vikings' own sensitive biology experiments.

The Vikings' landing sites had been carefully selected beforehand from the Mariner 9 reconnaissance, by a team led by Hal Masursky. Uppermost in their minds was the project's prime goal – the search for life – and the locations chosen were places where

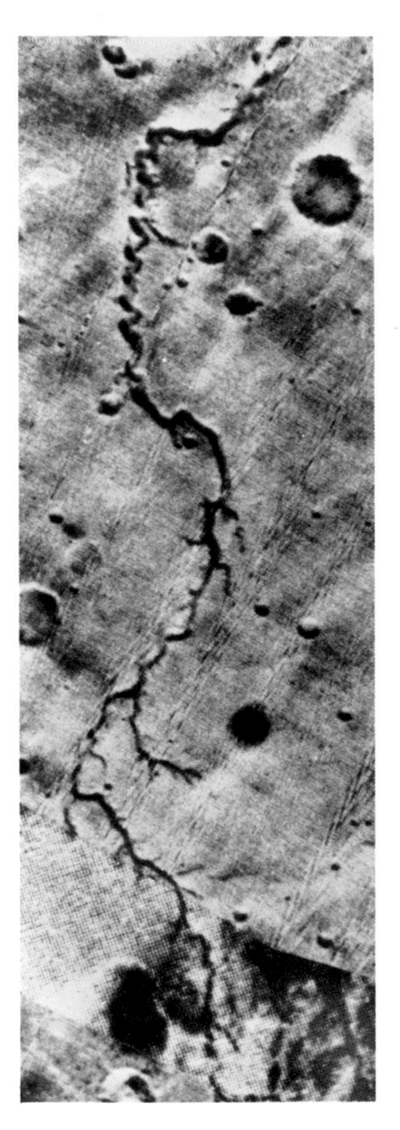

The first Mars-orbiting spacecraft, Mariner 9, sent back tantalising pictures of winding valleys which could only be the dried up beds of ancient rivers. This is the best example, a dry valley 350 miles (570km) long and 3 miles (5km) wide, with well-marked tributaries. If Mars once had running water, did it also have primitive life?

The huge volcano Olympus Mons towers above Mars's clouds, in this artwork based on photographs from the Viking Orbiters. This 15-mile (24km) high mountain is three times the height of Mount Everest – and on a planet only half the size of the Earth.

the terrain was smooth and safe for landing, but where the chances for finding life looked good, too. Viking 1's target was a spot about 20° north of Mars's equator, in the Chryse Planitia region, the 'Plain of Gold'. There, it appeared, vast ancient rivers had eroded a flat, potentially fertile vale.

When Viking 1 went into orbit on 19 June 1976, the orbiter's cameras began to reveal that the Chryse Planitia was even more eroded than originally thought. Although this implied that the site was even better from the biological standpoint – more water means more potential for life – it also meant that landing would prove more difficult. The 4 July touchdown – planned to mark the United States Bicentennial – was called off, and space scientists brought in the giant Arecibo radio telescope to test the roughness of Chryse by scanning the region with its powerful radar beams. Using a combination of these radar scans and the Viking's pictures from Mars orbit, the scientists selected a new site about 740km northwest of the original one. Just over two weeks later, Viking 1 made a flawless landing on the Martian sand dunes.

As Viking 1 landed, its companion craft Viking 2 was drawing rapidly closer to Mars. Its target landing area, close to 50° north latitude, was beginning to look even more unwelcoming than the Chryse site. When it entered Mars orbit on 7 August, the controllers at the Jet Propulsion Laboratory, Pasadena, encouraged by the success of Viking 1, decided that the spacecraft could seek out its own landing site by aerial photography. As a result, Viking 2 found a warmer area which was richer in water vapour than the spot originally selected. On 3 September 1976, the probe came down in the Utopia Planitia region (the 'Utopian Plains'), and landed on a gently sloping hillside.

The first task of both Viking landers was to survey the landing sites with their stereoscopic 'eyes', and return the pictures to Earth. Those early scans beautifully confirmed the views that NASA's space artists had been painting on all their publicity material. The red planet's rubble-strewn plains were indeed deep vermillion, and the almost airless sky a dark midnight blue. But every probe which takes colour pictures carries with it a 'colour wheel' – the equivalent of a decorator's shade card – and someone noticed that the colour wheels on the landers, as compared to their Earth-based twins, appeared out-of-balance. A spot of re-tuning followed. Then came the re-adjusted pictures, with the colours correct. The plains of Mars were, in fact, salmon-pink. Above, the sky was pink too – an effect of the fine dust suspended in the thin Martian atmosphere.

The Viking 1 Orbiter here reveals part of the immense canyon system on Mars, the Valles Marineris, which covers a total distance of nearly 3000 miles (4500km). The largest canyons are so wide and deep that the Swiss Alps could easily fit inside.

But there was nothing rosy about Mars's landscape. Drifts of fine Martian soil stretched for miles, as powdery as Antarctic snow (where some of the landers' tests had been carried out). Rocks and boulders of all sizes littered the view. Many were rough and volcanic-looking and some, like pumice, had small holes where gas had once bubbled through. Dominating the scene at Viking 1's Chryse site, however, was a

six-foot (2-metre) diameter boulder, bigger than all the others. Scientists at the US Geological Survey branch at Flagstaff, Arizona, who were responsible for compiling the lander photographs as huge mosaics, promptly nicknamed the rock 'Big Bertha'. But they had reckoned without the opposition from their feminist colleagues – and to this day, the rock is referred to as 'Big Joe'!

The Viking landers did far more than just take photographs. Over the four-year and six-year lifetimes, respectively, of Viking 2 and Viking 1, both landers returned weekly weather reports, analyses of the Martian air, and wind-speed measurements – as well as the thousands of pictures of Mars itself in all its moods. While they continued to function, the landers acted as automatic stations recording the day-to-day changes on another world. In fact, the long-serving Viking 1 lander was renamed 'The Thomas A. Mutch Memorial Station', after the brilliant young Viking teamleader who disappeared when climbing in the Himalayas.

But above all, the Viking landers will be remembered as the craft which searched for life on Mars. This was their main function; all other considerations were secondary. Each craft cost $500 million, and was equipped to see, sniff, touch and taste the Martian soil. The main problem the designers were faced with was: how would they recognize life if it *was* there?

The Viking 1 Lander surveys the red deserts of Mars. Parts of the spacecraft in the foreground include the meteorology boom (right) which monitors Mars's weather. The lander has dug trenches in the soil (visible to right of the boom), to test it for signs of life. In the middle distance is the large boulder nicknamed 'Big Joe'.

Although some of the scientists *were* guardedly optimistic about their chances of finding life, none of them expected to see it parading before the Vikings' cameras. If there was life, the chances were that it was microscopically small. But it would be a near-impossible task to send microscopes to Mars. So lateral thinking had to come in: even if the Vikings couldn't be designed to detect life directly, at least they could recognize it from its by-products and effects.

The Viking life-detection experiments were a masterpiece of ingenuity and – in the very limited space available – of miniaturization. The three main experiments were designed to fit into a container no bigger than an office wastebin, and each was to use no more than a thimbleful of Martian soil. They were cleverly designed – and had been tested in the deserts on Earth – to detect both animal and plant life.

From the moment the Viking 1 lander took up its first scoopful of Martian soil, a controversy began to rage as to whether Mars has life – or not. For the first few days, however, there appeared to be no doubt. The scientists were ecstatic, convinced they had found Martian life.

The first experiment was designed to test for both animal or plant life, and relied on the fact that if you give an animal or plant something to eat, it sooner or later gives off gas. The Viking scientists were astonished that almost immediately after 'feeding' their soil sample with a rich 'broth', copious amounts of carbon dioxide began to be given off. But after while, the levels fell off. This wasn't what was expected. If there *had* been life in the soil, it should have been busily reproducing and generating even more gas. What was going on?

The second experiment also gave confusing results. It was designed to test specifically for the presence of animal life – little 'bugs' in the soil. The bugs were fed with what should have been a delicious soup and, almost instantly, the gas levels rose to a peak. But not long after, they dropped.

Experiment number three was a test for plant life. In order to live and grow, plants on Earth take in sunlight and carbon dioxide. Would the Martian soil absorb carbon dioxide if it was fed with it? The answer was yes; but it was still 'yes' after the soil sample had been baked to a temperature when all plant life would have been burnt to a crisp.

Understandably, the Viking scientists were baffled by the puzzling behaviour of the Martian soil. Was there life there or not? 'All the signs suggest that life exists on Mars, but we can't find any bodies!' despaired Gerry Soffen, the Viking Project's chief scientist.

The test that now seems to have proved conclusive was the simplest of all. This experiment merely involved 'baking' lumps of the Martian soil and testing for the kinds of gases given off. Organic material – that containing carbon, as all lifeforms must – gives off a very special aroma when it is heated, as anyone fond of their Sunday roast knows! That unique aroma is due to organic compounds in the food. But the cooked Martian soil gave off no characteristic smells. In fact, there were no indications that it contained any carbon compounds at all, let alone those we would expect to find in carbon-rich living cells.

And so, in spite of centuries of hope and speculation, Mars really does seem to be lifeless. Why? After conditions first appeared so promising for life, why hasn't it developed? One of the reasons is that Mars appears to be gripped by a severe ice age. The temperature at the equator on Mars never rises above 20°C, even on the hottest day of the year. At night, it is likely to fall below minus 120°C.

We know that this ice age cannot have lasted for ever, because rivers once flowed on Mars. They must have flowed for more than just a short time, too, because their valleys are deeply incised, not the result of a single flash flood. But it all happened such a long time ago. We only find traces of the river beds in Mars's very oldest regions, its uplands. Since then, the planet has just been too cold, its meagre water supplies frozen into its icy polar caps, or as permafrost in the Martian soil. That's why the iron-rich surface of Mars is the colour it is, rusted red by its locked-up water.

There's another factor. Not only is Mars's atmosphere now far too thin to support life, but there is hardly any oxygen. On Earth, oxygen has a dual function. It is essential for all animal life; we need to breathe oxygen in order to survive. But oxygen has a protective role to play, too. The ozone layer of oxygen surrounding our planet absorbs most of the Sun's harmful ultraviolet radiation, which would otherwise burn and destroy all living things. Mars has no ozone layer, so the Sun's ultraviolet rays get through, and as a result, the Martian soil has been rendered completely sterile. Another effect of the ultraviolet bombardment has been to give the soil strange and unexpected chemical properties. The Viking experiments discovered this for themselves at first hand, and it was these unusual properties that scientists first interpreted as signs of life.

Sunset on Mars – as seen by the Viking 1 Lander in August 1976. (The distinct stripes in the sunset glow were caused by computer processing of the image.)

Huge volcanoes rise from the Tharsis ridge, which is a large swelling in Mars's surface. The 'ground level' on Tharsis is six miles higher than the rest of Mars, and the swelling of the surface here has stretched the flanks of Tharsis, causing the cracks that are visible at the right.

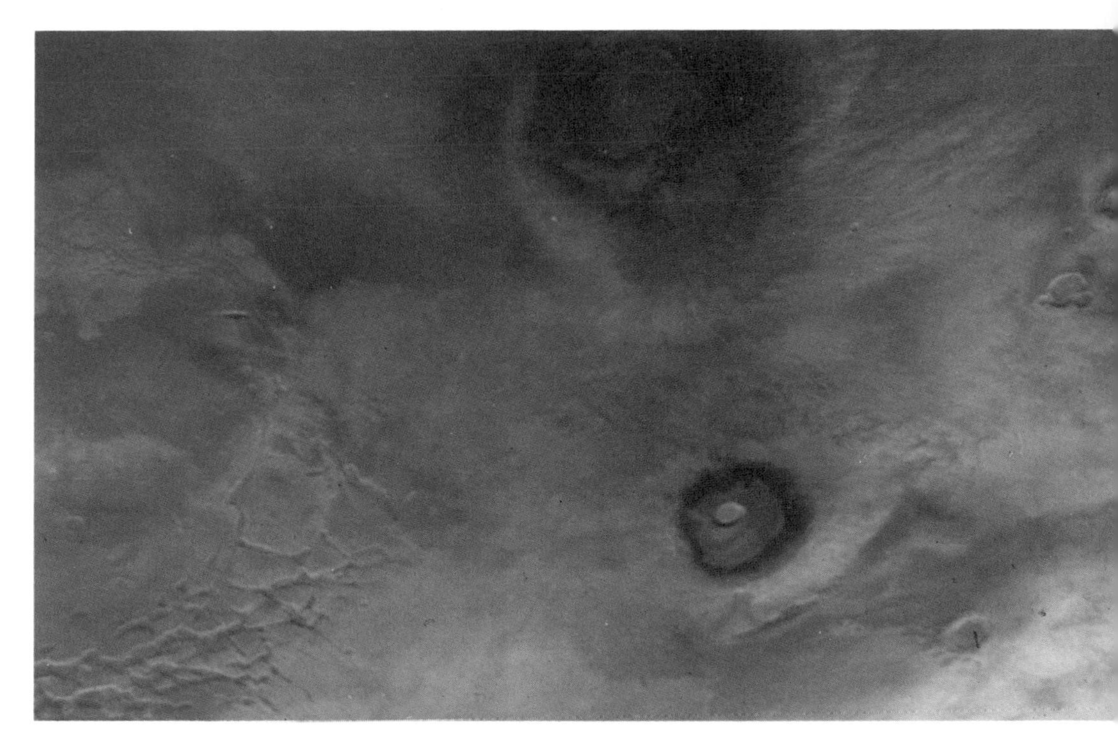

Some of the youngest volcanoes on Mars are found in a geologically active region near the edge of the Tharsis ridge. The pattern of parallel cracks at the left is still something of a geological mystery.

All this may be the case *now*, but astronomers can't help thinking back. Mars must have been far more welcoming in the past. To support deep running rivers, Mars's atmosphere must have been a lot thicker. The whole planet would have been much warmer. In certain places – although not where we have looked – very primitive life may have got started. So why and when did things change?

One key to the change may involve the huge Martian volcanoes on the Tharsis Plateau. In their heyday, their eruptions could have produced sufficient quantities of gas and water to create a substantial atmosphere and an extensive river system on Mars. Now, though they appear extinct, they could be 'only resting'. Bizarre though it may sound, geologists collecting meteorites in Antarctica have found a number of red-coloured stones which appear to have crystallized only 600 million years ago. This is much too young to have originated in the asteroid belt, but the meteorites have certainly come from outside the Earth. One idea is that they were born in lava flows on Mars, and were subsequently 'splashed' in the direction of Earth by the impact of a huge meteorite on the planet itself. If true, it means that the volcanoes on Mars were active in geologically recent times.

The Viking probes have, predictably, raised many more questions than they have answered. Quite apart from the life aspects, Mars has turned out to be the most fascinating and welcoming of all the planets. Its chasms, river valleys, volcanoes and polar caps are all crying out for more detailed investigation.

The Americans are planning to return with their orbiting Mars Observer craft, due for launch in 1990 or a little after. The unmanned instrument-packed probe will study the surface and environment of Mars in unprecedented detail from an orbit only 220 miles (350km) above the surface of the planet. The information it sends back, after a year or more of reconnaissance, will form the basis for the next mission in 1996–2000 – the Mars Rover.

At the University of Arizona at Tucson, a bunch of graduate students are building a mock-up Mars Rover in their spare time. It's the only student project at the moment to be funded by a NASA grant, but the students themselves have no idea if their prototype stands any chance of actually getting to Mars. The Tucson 'Mars Ball' is nothing more than a mobile air-bag. The students' first attempt was four feet (1½ metres) high, powered by a leafblower, and cost just $200. But the reality – if it works – will be a lot more impressive. Their plan is to build an enormous roving vehicle with two inflatable wheels twenty feet (6 metres) high, and an instrument package (containing TV cameras and experiment platform) suspended safely between them. In flight, each wheel – made of sixteen separate pie sectors of plastic material – will be deflated and flat. But once on Mars, the Rover will be able to drive itself over the barren plains by sequentially inflating and then deflating the sectors of its wheels. The Rover will cope with rocks up to three feet high, and a radar detector will warn of holes. At present, the prototype can go up and down slopes of 30° slant, while the real thing will be able to take a tumble without damage. Hopefully, the tyres will even cope with a puncture – they'll be double-chambered.

Far-fetched though it might sound, the Mars Rovers project has the enthusiastic approval of many scientists. Instead of being limited to a 'safe' site, like the Viking landers, it could investigate the interesting – and predictably, more dangerous – terrain on Mars where complex geological processes have been at work. One very risky area which planetary expert Hal Masursky is particularly keen to study is that of the slopes of Mars's super-grand canyon, the Valles Marineris. Here he wants to look at the layered rocks deposited by rivers millions of years ago – for they may contain evidence of primitive life nipped in the bud.

One of the goals of the Mars Rover mission would be to drill samples from a range of environments on Mars and return them for analysis to Earth – or rather, to the US space station that NASA hope to have permanently in orbit about Earth by then. With the knowledge so gained, the way ahead would then be clear for the attempt on Mars by a manned mission.

A striking false-colour picture of Mars's largest volcano, Olympus Mons, shows slight differences in the lava flows that run down its flanks. Processing of this kind can show up different kinds of geological terrain, and so pinpoint areas of especial interest for the future landing of Mars Rovers and manned expeditions to the planet.

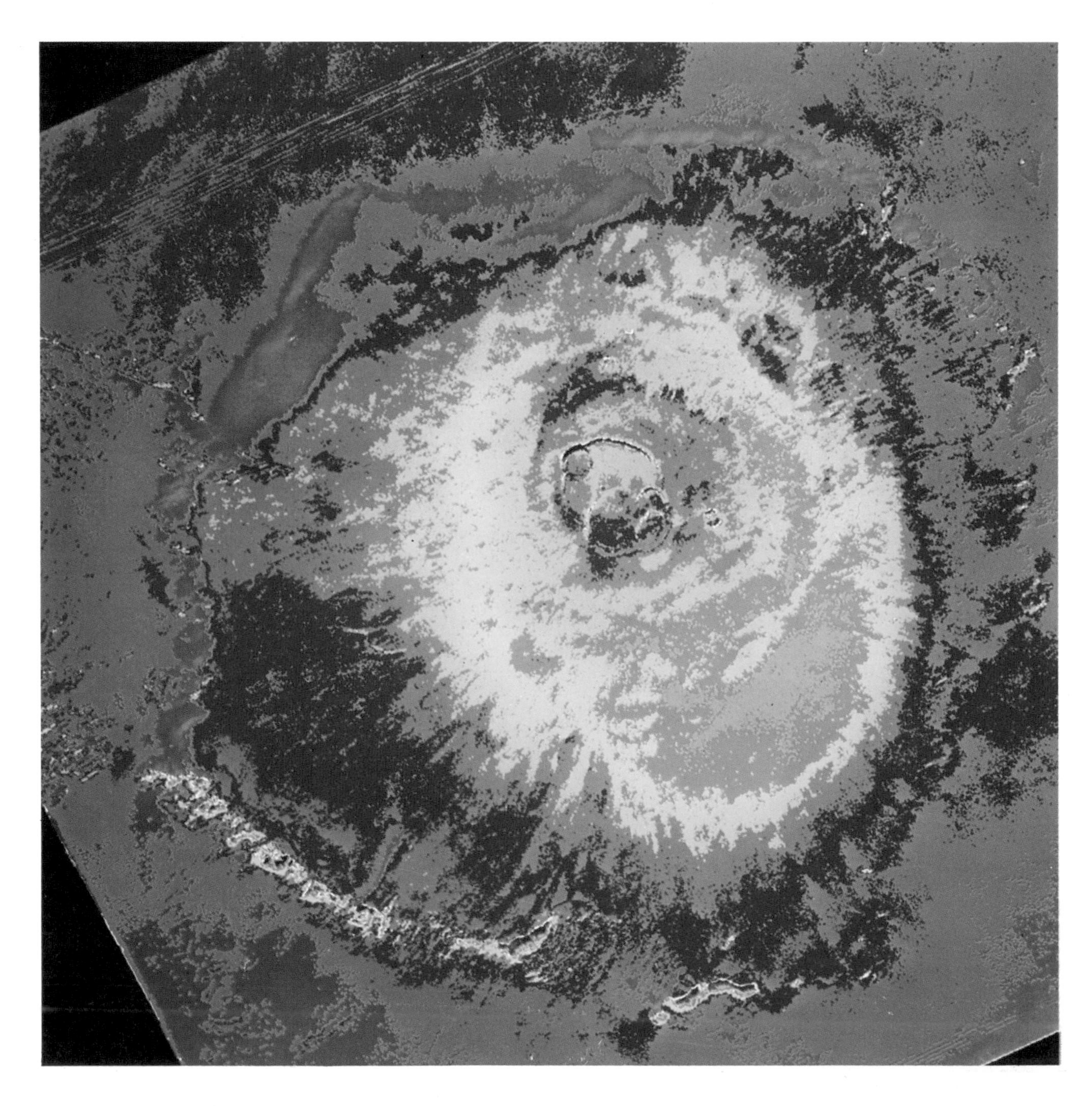

But the American plans for Mars sound positively tame compared to what the Russians have in mind. In a new and encouraging spirit of openness and co-operation, Soviet space scientists have just unveiled their latest planetary exploration programme to delegates at a recent US conference. One US scientist said bemusedly: 'It was a real eye-opener. The detail and the candour with which things were discussed . . . was uniformly impressive to everyone present. We . . . are feeling kind of foolish, because now they have left, and we are still coming up with important questions about what they said – while during the briefing we sat there marvelling at how much they were telling us in the first place!'

Most of all, the US scientists were impressed by the Soviets' plans for Mars. This far, the Russians have had very little success with the planet. Their seven Mars probes were all failures in one way or another, prompting a US scientist to label their Mars programme 'a catastrophe'. One of their most spectacular failures was to send a lander to Mars inside a ball-shaped container – it rolled into a big crater and was never heard of again!

But those days have gone for good. Very shortly, the Russians will launch a Mars mission which is so innovative that it 'boggles the mind', according to one American astronomer. In 1988, two laser-equipped spacecraft will be sent into Mars orbit. One will then manoeuvre to within 200 feet (60 metres) of Mars's larger moon, Phobos, and fire laser beams at its surface to vapourize a sample for analysis. Afterwards, it will drop a lander vehicle designed to hop across Phobos's surface, taking pictures and returning other data. Suppose it crashes? If it does, the other vehicle will attempt the same manoeuvre; but if not, the second vehicle will go on to repeat the whole exercise with Mars's smaller moon, Deimos. 'That flight is quite a big deal, and if it was one of our own we would be rather proud of it,' comments an American space official.

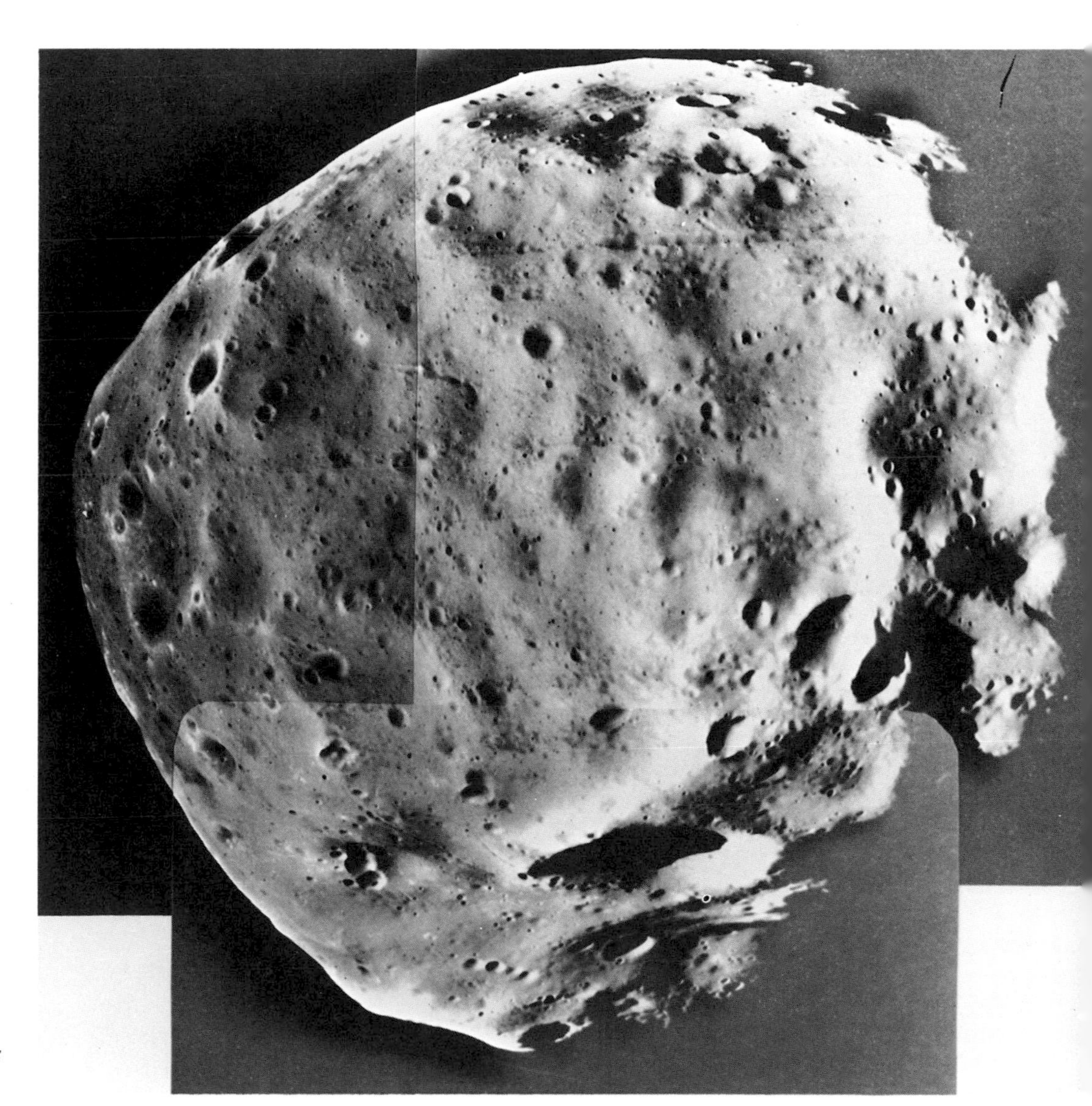

The battered surface of Mars's larger moon, Phobos, is clearly revealed in this picture from the Viking 1 Orbiter. Future orbiters will blast pieces off Phobos with a laser, and land small probes on its surface.

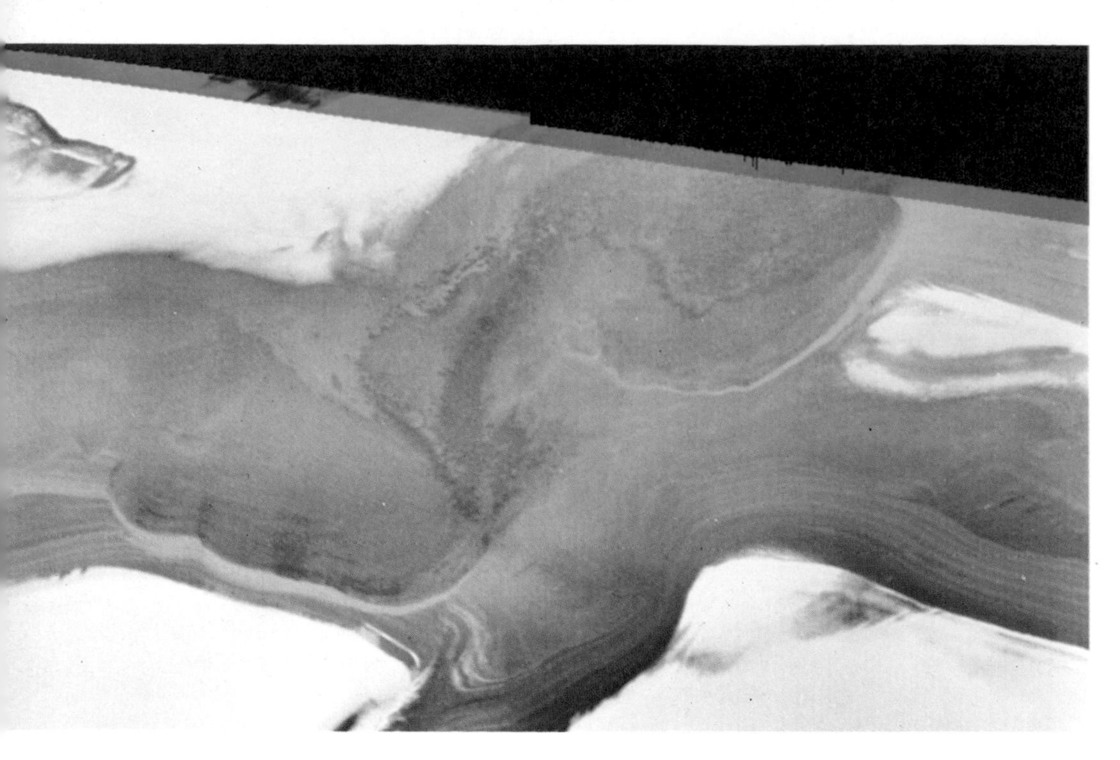

At the edge of Mars's polar caps, the icy surface is cut by deep valleys that reveal layers of sediment below. The regular succession of sediments probably means that this cold planet has its own version of the ice ages that have affected the Earth. By examining these layers with future Mars expeditions, we can hope to learn more about the fluctuations in the Earth's climate.

Where next? The Americans can only speculate. Some maintain that the Russians are concentrating on the Martian moons in preparation for a later manned mission. There, in practically zero gravity and no atmosphere, the engineering difficulties would be far less formidable than in attempting a landing on Mars itself. But others disagree – why should the Russians go through all the hassle of the nine-month journey to Mars and then not land on the planet itself?

Whatever the plans are, both sides now seem to be talking about co-operation. Hal Masursky notes that the Phobos laser probe and the Mars Observer have four instruments in common; he wants to cross-calibrate them before the Soviet launch. And a US spokesman, on hearing of the new Russian programme, remarked: 'We could suddenly find ourselves in a space race again, if we choose to be, or conversely we may choose to sit down together and chart joint programmes in which we could make real progress . . . We could now divide up the big planetary goals and do them twice as fast because we could bring twice the resources to bear.'

Mars may be famous as the 'Red Planet', and perhaps the odds are now on that Russia, appropriately enough, will land its cosmonauts there first. But let us hope that this new spirit of co-operation endures. If it does continue, it might be a multinational team which greets the twenty-first century from Mars's not quite red, but salmon-pink slopes.

CHAPTER SEVEN

JUPITER

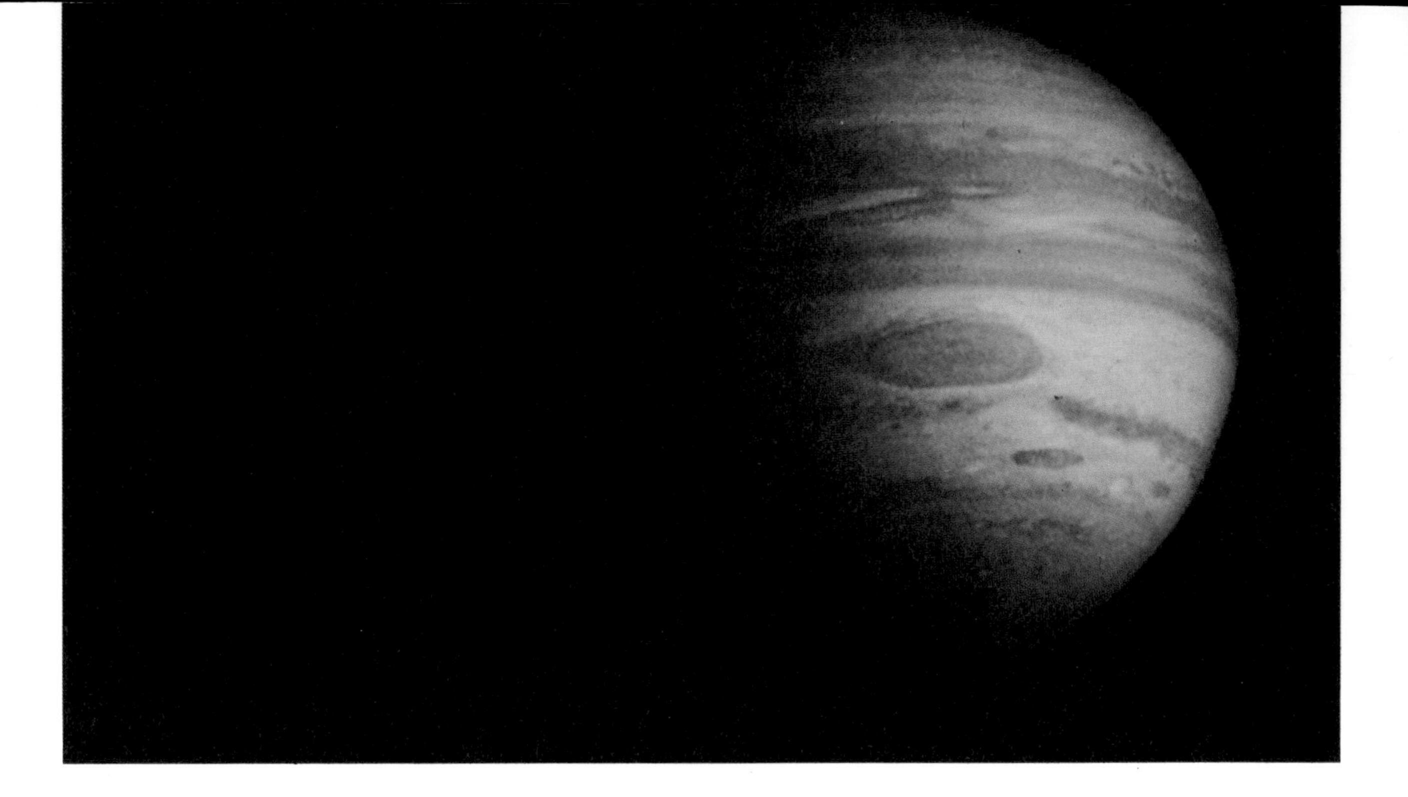

When the astronomers of ancient Babylon and Greece named the most stately of the planets for the king of the gods, they chose more wisely than they could have imagined. Their observations showed that Jupiter moves around the sky at a regal pace, taking 12 years to complete one circuit. Its constant yellowish light outshines all the planets bar brilliant Venus; and while Venus stays close to the Sun and is seen only in the twilight, low down on the horizon, Jupiter can hang high overhead, brilliant against the darkness of the night skies. It is by far the largest planet. Jupiter's great globe is as wide as eleven Earths, and if Jupiter were hollow it could contain over a thousand planets like ours. It is so massive that Jupiter outweighs all the other planets put together twice over.

Previous page: *Jupiter's Great Red Spot dominates this picture of the giant planet, taken by the Voyager 2 spacecraft in 1979. The spot, a huge eddy pattern in Jupiter's clouds, is three times as large as the Earth.*

Jupiter is so large that you do not even need a telescope to see its globe. Look through an ordinary pair of binoculars, firmly supported to keep the view steady, and you can make out Jupiter as a distant oval disc. You will also see some faint, star-like specks of light to either side of the planet. These are the four largest of Jupiter's family of sixteen moons.

A small telescope shows the pattern of the clouds that swirl over Jupiter's globe. Light and dark bands run across the planet, parallel to its equator, and their appearance changes gradually over the months and years, reflecting changes in Jupiter's 'weather'. Nestled between the light 'zones' and darker 'belts', you will sometimes spot some small oval shapes, the tops of swirling cloud systems. Most of the ovals are white, and last only a few years. But one of them stands out, for its size, its colour and its longevity: the Great Red Spot.

In the 1660s, when telescopes became powerful enough to show details on Jupiter, the British scientist Robert Hooke reported 'a spot in the largest of the three observed belts of Jupiter . . . its diameter is one-tenth of Jupiter'. An Italian, Giovanni Cassini, saw the spot at the same time, and astronomers continued to report similar spots at intervals for the next two centuries. In 1878, astronomers were surprised to find a large oval, brick-red in colour, dominating Jupiter's appearance. This 'Great Red Spot' faded after a few years, but astronomers have been able to see it more or less distinctly ever since, although rarely with such a prominent colour. We can't be certain, but it's very likely that the spots seen since the seventeenth century were all the Great Red Spot, at times when it was particularly prominent – so this peculiar red cloud pattern has lasted more than three centuries.

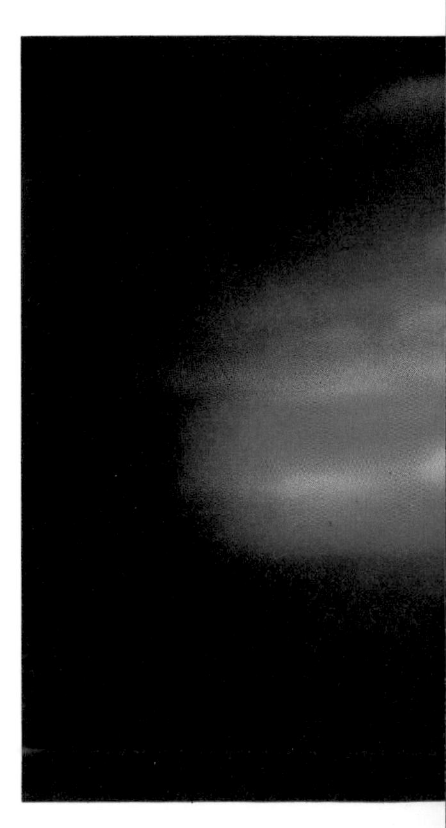

The Pioneer spaceprobes took the first close-up pictures of Jupiter. When Pioneer 11 sent back this view in 1979, the Great Red Spot was intensely coloured and surrounded by featureless white clouds. Jupiter's restless atmosphere broke up these clouds by the time that the Voyagers passed by, five years later.

By the 1970s, astronomers were looking forward to the space probes that would examine Jupiter in detail, revealing the true motions of the clouds and spots that are tantalizingly glimpsed from the Earth. But the centuries of ground-based observations had already led to a good broad-brush picture of what Jupiter is like, inside as well as out.

Early on, astronomers had realized that Jupiter could not be made of rock like the Earth. The argument went like this: Jupiter is big enough to contain 1300 Earths, so if it were made of the same kind of material, it would weigh 1300 times as much as our planet. The motion of Jupiter's moons tells us the strength of the planet's gravity and hence how massive ('heavy') Jupiter is: it turns out to weigh only 318 Earths. This means that Jupiter's matter must be considerably less dense than the rocks that constitute our planet. And when we remember that Jupiter's gravity must be squeezing the material in its centre to a density much higher than usual, it turns out that Jupiter should consist almost entirely of gases.

By analysing the sunlight reflected from Jupiter, astronomers have found that it is made mainly of the light gases hydrogen and helium (in the proportion nine parts of hydrogen to one of helium). These gases are relatively uncommon on the Earth, but in this instance it is our planet that is the oddity. Most of the matter in the Universe consists principally of hydrogen and helium: for example, the stars (including the Sun) and the very thin gas between the stars. The Earth's atmosphere may once have held more of these gases, but its gravity is not strong enough to hold on to either hydrogen or helium, and they have escaped to space. But Jupiter's stronger gravity is well able to retain the gases that make up its enormous bulk.

In a false-colour 'heat picture' of Jupiter, the cooler regions appear blue and green and the planet's warmer parts are reddish. The broad pink stripe is a gap between Jupiter's higher layers of cloud; here we are seeing down to lower cloud decks which are warmed by heat coming up from Jupiter's hot core.

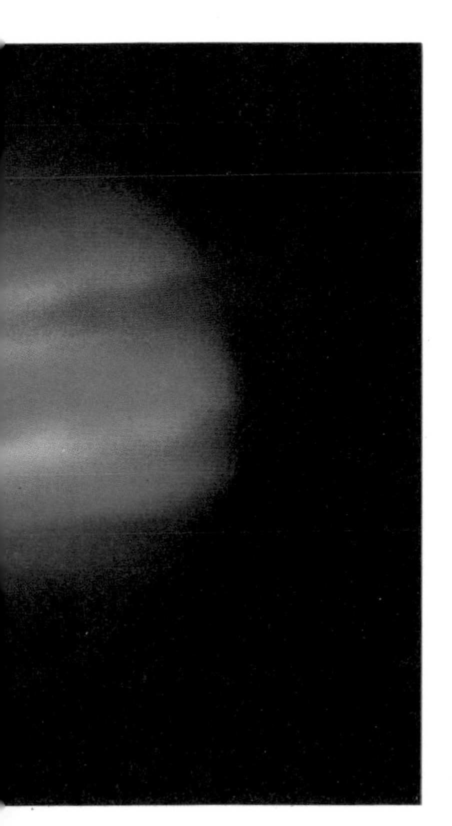

When we look at Jupiter, then, we are not seeing a solid surface, like that of the Earth or the Moon. Nor are we seeing just a layer of cloud that covers a solid surface, as we see on Venus. Jupiter's clouds cover an immense depth of hydrogen and helium that constitute most of the planet. On Earth, we are familiar with different clouds; in an aircraft, we often fly above layers of water-drop clouds, and below the high 'mares-tails' of the icy cirrus clouds. Jupiter's clouds are more exciting. There are several different substances in its atmosphere that can condense into clouds, at different levels and displaying different colours.

The highest clouds are made of ammonia, a substance that we know on Earth as a pungent gas, but which on Jupiter is frozen into icy crystals. The ammonia clouds are blindingly white and stretch all the way around the planet at certain latitudes, comprising the great white zones that are visible with even a small telescope.

Between the zones, we look down to a deeper layer of cloud. These are probably made of a compound of ammonia and another substance we meet in school chemistry labs as an unpleasant gas, hydrogen sulphide, which smells like rotten eggs. The latter releases sulphur and sulphur compounds that tinge the clouds in shades of orange and brown, creating the effect of dark belts between the higher altitude white zones. The strong colour of the Great Red Spot is different again. The top of the spot is at a high altitude, above even the ammonia clouds, and its red colour is probably due to phosphorus (the substance used for the heads of 'strike-anywhere' matches) dredged up from far below.

Occasional small holes in the deeper clouds of the belts let us see through to a still deeper layer of cloud, possibly made up of water drops, like the Earth's familiar clouds. But that is as far as we can see into Jupiter; and it takes us only 50 miles (80km) into the giant planet.

But we can make a journey to the centre of Jupiter, in imagination at least. Bill Hubbard, of the University of Arizona, has followed other leading theorists of the past by amassing all the latest clues that tell us about Jupiter's deep interior. The detective-work involves piecing together pieces of astronomical information (such as Jupiter's low density) with information from laboratories (including the effect of compressing hydrogen) and the latest theories of matter at pressures too high to achieve in the laboratory.

Our journey into Jupiter takes us very quickly through the layers of cloud. In proportion to Jupiter's size, they are no thicker than the skin of an apple. Below, we come into a hazy atmosphere that extends down and down. In its depths the hydrogen-

The big dishes of the Very Large Array, near Socorro in New Mexico, form a very sensitive radio 'eye' on the Universe.

helium gas is denser, compressed by the weight of the gas above. Pressure is higher, as in the depths of the Earth's oceans. When we have travelled only a fraction of the way to Jupiter's core, we find that the 'gas' has become so thick and dense that it is behaving like a liquid. Below this level, we find that it is indeed more like travelling in a submarine than an airplane – although we never crossed any definite surface to mark the difference between the 'atmosphere' and the 'oceans' of Jupiter.

Further down still, about one-third the way to Jupiter's core, something even more startling happens. The ocean around us changes from being a cloudy but colourless liquid, to an ocean of liquid metal, shining silver like the mercury in a thermometer. This ocean is still composed of the same hydrogen-helium mixture, but this deep within Jupiter the pressure is so high that the hydrogen looks and behaves like a metal.

At the end of our journey, right at Jupiter's centre, we find a core made of rock. Hubbard's calculations tell us that the core is probably about fifteen times heavier than the Earth, but squeezed into a region only about twice the size of our planet.

Although we have not actually probed Jupiter's centre in this way, there's another good reason to believe that the planet really is composed largely of an ocean of hydrogen metal. In 1955, radio astronomers Bernard Burke and Kenneth Franklin picked up strange bursts of radio static coming from the sky. To their surprise, they traced the broadcasts to the planet Jupiter. The radio waves come from a region around Jupiter that looks like empty space in an ordinary telescope. But the radio telescope's view shows us that it is filled with a powerful magnetic field, extending a long way from the planet itself. This vast magnetic region, Jupiter's magnetosphere, traps electrically charged particles that whirl about in a vast cloud, like the Earth's Van Allen belts; and these particles broadcast the radio waves.

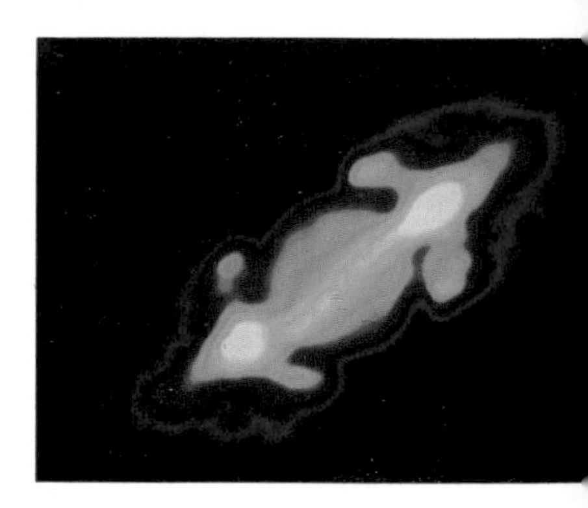

Jupiter's Van Allen belts show up when it is 'seen' by radio telescopes. In this view, from the Very Large Array, the planet is the central green oval; the Van Allen belts form the white and blue band girdling Jupiter.

We are familiar with the Earth's magnetism, which lines up compass needles north-south. The Earth's magnetism is generated in its core of liquid iron, where the planet's rotation generates electric currents. Jupiter's magnetism is 20 000 times stronger than the Earth's, and so it must have a liquid-metal core to act as a magnetic dynamo. In Jupiter's case, the liquid metal must be the hydrogen that makes up most of its bulk.

Jupiter certainly spins fast enough to drive a magnetic dynamo. Despite its vast size, the planet turns more than twice as quickly as the Earth, so Jupiter's 'day' lasts only 9 hours 55 minutes. At such a terrific rotation rate, Jupiter's equatorial regions are flung outwards, to make the planet bulge at the equator. A small telescope, or even a good pair of binoculars, will show how Jupiter is distended at the equator, or – as it is usually described – flattened at the poles.

In recent years, measurements of Jupiter's heat have revealed something even more bizarre about the planet's interior. It has slight stirrings of being a star.

Astronomers expect all planets to be emitting heat radiation – infra-red – simply because they are warmed by the Sun. But when they turned infra-red telescopes to Jupiter, astronomers found that it emits considerably more heat than it receives from the Sun. Infra-red pictures of Jupiter show which parts of the body are abnormally hot. They show that the warmest parts of Jupiter are the lower cloud regions of the dark belts; and we see particularly brilliant 'hot spots' where breaks in these clouds reveal layers even further down. Somehow, Jupiter must be generating heat deep within its core.

The apparently empty space around Jupiter is in fact filled with electrons and magnetic field, making up a huge magnetosphere. This diagram shows the magnetosphere (purple) moulded by the force of particles (right) from the Sun; the yellow disc and orange ring represent gases circling Jupiter. The planet itself, within the ring, is too small to show on this scale: its invisible magnetosphere is much larger than the Sun.

Jupiter's heat, in fact, seems to be a reminder of its birth. The planet must have condensed from a great cloud of hydrogen and helium, and as this gas was compressed by the gravity of the forming planet, it must have given out heat – just as the air in a bicycle pump becomes hot if you squeeze in the piston with your finger over the outlet. Jupiter has not yet reached its final size. The planet is probably still shrinking – and its heat output could result from a shrinkage of only a fraction of an inch (a few millimetres) per year!

When the Sun – and other stars – formed, it too began to generate heat by contracting in on itself. The Sun is a thousand times more massive than Jupiter, and the temperature at its centre reached ten million °C; at this point, fierce nuclear reactions began, and these reactions still continue to provide the Sun's immense output of

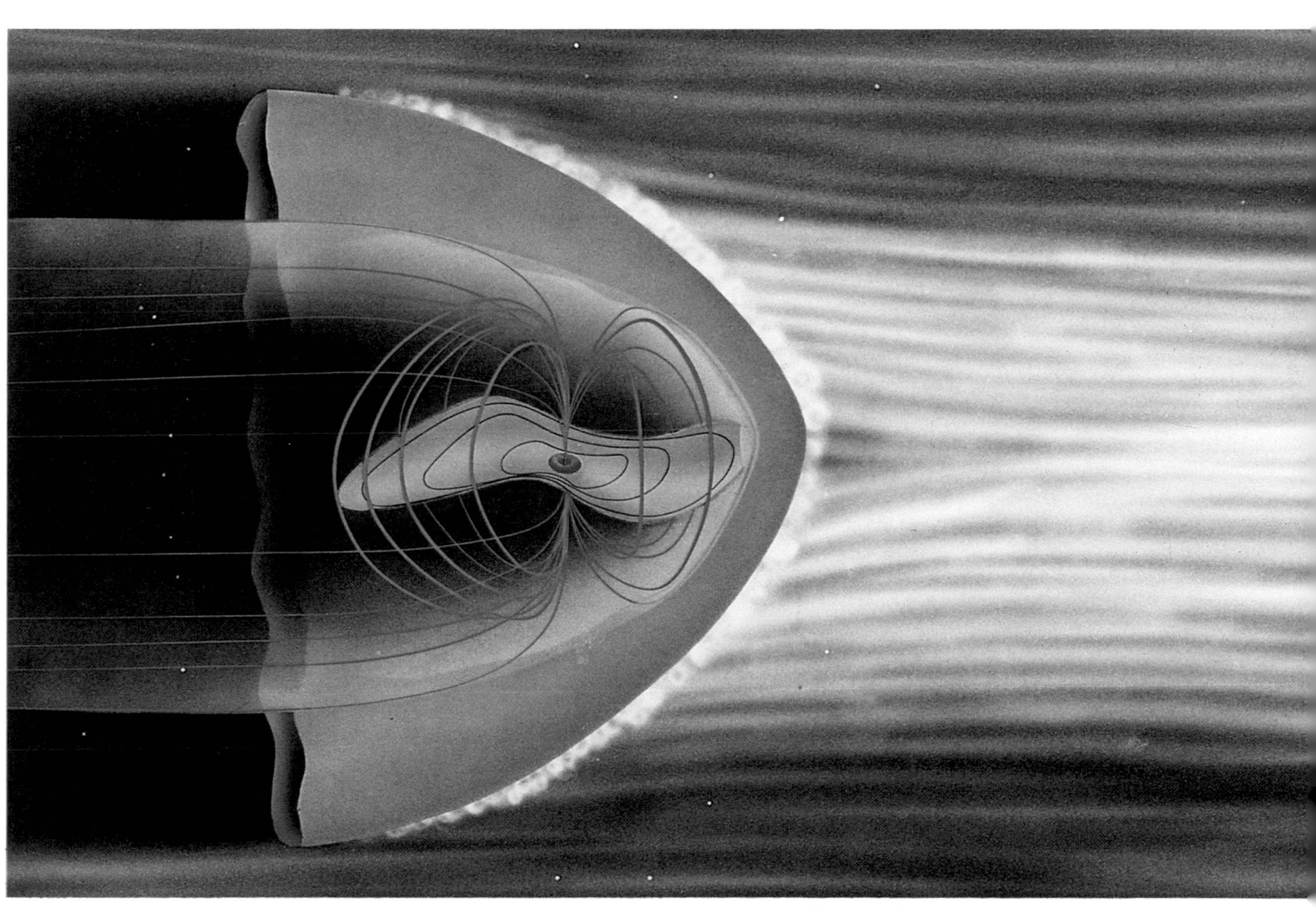

energy. Jupiter's centre never became so hot – the temperature at its core is probably about 30 000°C – and so it could never switch on the nuclear reactions that would have made it shine as a star.

Because Jupiter started off as on the same path as the Sun, however, some astronomers see it as a 'failed star'. Frank Low and Don McCarthy have used an infra-red telescope to discover a rather heavier planet going around another star (see Chapter 12). They propose that this object (VB8B) and Jupiter are a new kind of beast in the astronomical zoo. Too lightweight to be stars, but too heavy to be the completely cold bodies that we normally call planets, they are 'brown dwarfs'.

Even if we choose to call Jupiter a brown dwarf, rather than a planet, astronomers agree that there is little chance that it could be 'switched on' to shine as a star – as happens in Arthur C. Clarke's film *2010*. But many astronomers would admit that brown dwarf Jupiter resembles a star in another way: its family of sixteen moons is in many ways like a solar system in miniature, with Jupiter rather than the Sun in control. Space probes sent to Jupiter are designed to study the moons as much as to survey Jupiter itself.

When astronomers first began to study Jupiter, the discovery of the four brightest moons was far more important than their observations of Jupiter. They were pivotal in swinging man's idea of the cosmos from an Earth-centred view to an acceptance of the Sun as the centre of the planetary system.

On the night of 7 January 1610, Galileo turned his recently improved telescope – still no better than modern binoculars – towards Jupiter. His 'very excellent instrument' revealed three little stars, very near to Jupiter, and lying in a line. The next evening, the stars were even closer together, and had changed their positions. On 13 January, he discovered a fourth 'star'. All four were constantly in motion, shifting back and forth as the nights went by. Galileo concluded that there are 'four, erratic sidereal bodies performing their revolutions about Jupiter'. He call them 'sidereal' (starlike) because they appeared as points of light in his telescopes; in reality they were the four largest of Jupiter's moons, or satellites.

Galileo was delighted with his discovery, for it gave him strong ammunition in his ongoing arguments with the majority of scientists, who still believed that the Earth was the centre of the Universe. The Earth-centred view was very simple to believe: the Earth is stationary, and everything goes around it. Galileo sided with the new opposition, proposed by Nicolas Copernicus in 1543, who said that the five planets and the Earth travel around the Sun. But no one could dispute the fact that the Moon goes around the Earth. Copernicus and Galileo had to have two 'centres of motions': the Earth at the centre of the Moon's orbit; and the Sun at the centre of the planetary system. Even worse, critics argued that if the Earth were moving around the Sun, then the Moon should be left behind.

Galileo's discovery of Jupiter's moons countered that criticism very neatly. Clearly, it was quite possible for a planet to be a 'centre of motion' in its own right, and if Jupiter could keep four moons in tow as it moved, the moving Earth should have no problem holding on to just one Moon. Although Galileo's observations of the Moon, and of Venus, also supported a Sun-centred Universe, Jupiter's moons provided the strongest evidence.

Astronomers have honoured this achievement by calling these four moons 'the Galilean satellites' of Jupiter. In fact, Galileo was not the first to see these tiny specks of light through a telescope. The German astronomer Simon Marius had spotted them a few days earlier. But Marius did not follow up his discovery with Galileo's painstaking care, nor did he comprehend how these small 'stars' could change man's conception of the heavens.

As some kind of recompense, however, the four moons now bear names suggested by Marius, commemorating four of the amorous conquests of Jupiter the god: Io, Europa, Ganymede and Callisto.

The mythical Io and Callisto were both transformed into animals – a white heifer and a bear respectively – by Jupiter, in order to protect them from the jealousy of his wife, Juno. Despite this ruse, Juno sent a gadfly that drove the heifer Io to Egypt; and she had Callisto hunted down by her own son – although fortunately Jupiter intervened and swung her up into the sky, where Callisto is seen in another, much older, astronomical context as the constellation of the Great Bear. In Europa's case, Jupiter turned himself into a white bull in order to abduct her from the seashore where she

22 OBSERVAT. SIDEREAE

que talis poſitio. Media Stella oriẽtali quam p

Ori. * ** ✡ * Occ.

xima min. tantum ſec. 20. elongabatur ab illa

a linea recta per extremas, & Iouem produ

paululum verſus auſtrum declinabat.

Die 18. hora 0. min. 20. ab occaſu, talis fui

ſpectus. Erat Stella orientalis maior occiden

Ori. * ✡ * Occ

li, & a Ioue diſtans min. pr. 8. Occidentalis v

a Ioue aberat min. 10.

Die 19. hora noctis ſecunda talis fuit Stell

coordinatio: erant nempe ſecundum rectam

Ori. * ✡ * * Occ

neam ad vnguem tres cum Ioue Stellæ: Orien

lis vna a Ioue diſtans min. pr. 6. inter Iouem,

primam ſequentẽ occidentalem, mediabat m

5. interſtitium: hæc autem ab occidentalior

berat min. 4. Anceps eram tunc, nunquid i

orientalem Stellam, & Iouem Stellula media

verum Ioui quam proxima, adeo vt illum

tangeret; At hora quinta hanc manifeſte

medium iam inter Iouem, & orientalem Stel

locum exquiſite occupantem, ita vt talis fu

Ori. * • ✡ * * Occ

configuratio. Stella inſuper nouiſſime conſp

admodum exigua fuit; veruntamen hora

reliquis magnitudine fere fuit æqualis.

Die 20. hora 1. min. 15. conſtitutio conſin

viſa eſt. Aderant tres Stellulæ adeo exiguæ, v

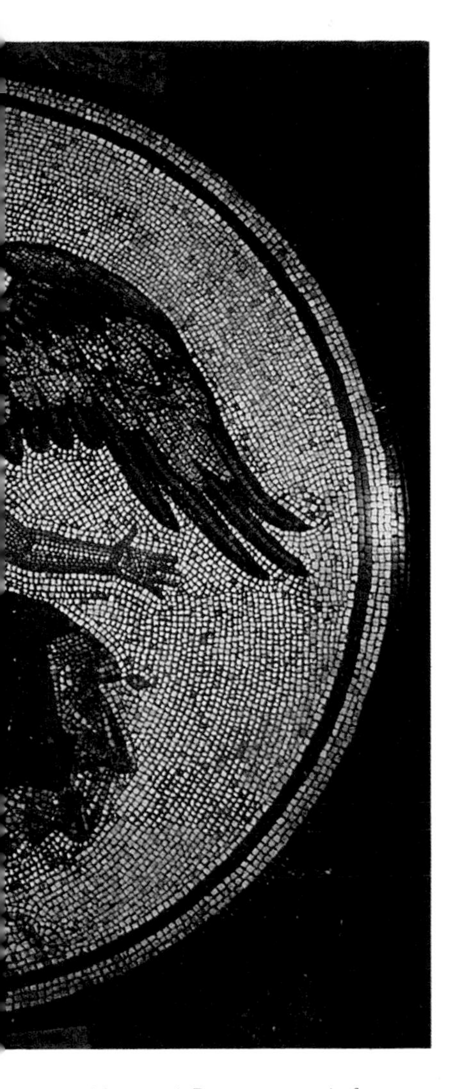

was playing with her maidens. Ganymede was the most beautiful of mortal youths, and Jupiter – in the shape of an eagle – carried him off to Olympus to be the cupbearer to the gods, sending his father a pair of horses in compensation.

In his writings, Galileo referred to the four satellites as 'planets' – meaning 'wandering stars'. Although we would now call them moons or satellites of Jupiter, in one sense Galileo's word is more fitting, for these worlds are every bit as large as the smaller planets. Jupiter's other satellites are much smaller, and fainter. During the seventeenth, eighteenth and nineteenth centuries, astronomers were finding more and more moons for Saturn and Mars, and for the newly-discovered planets Uranus and Neptune; but despite intensive searches, Jupiter seemed to have only four. Then, in 1892, one of the keenest-sighted astronomers in history, E. E. Barnard, began to use the great 36 inch (91cm) telescope of the Lick Observatory. On 9 September he glimpsed a faint point of light – a new satellite, closer in even than Io's orbit around Jupiter. He named the satellite after Jupiter's nurse, the goat Amalthea. Amalthea is almost ten-thousand times fainter than Ganymede; and the rest of the family of small moons known before the space probe era were only caught when astronomers could use photographic plates to record their faint images.

Eight of these small irregular moons orbit Jupiter outside the paths of the Galilean satellites. The first four go round Jupiter in the same direction as the Galilean moons, and are probably fragments of rock and ice left over from Jupiter's birth. The outermost four moons, oddly enough, go round the other way. They may be fragments of rock from the asteroid belt between Jupiter and Mars, drawn into orbit around Jupiter by the giant planet's gravity.

The smaller moons may carry some clues to Jupiter's early days; but astronomers have always been more fascinated by the great Galilean moons, especially Io, the closest to Jupiter. These worlds are so far away from us that telescopes on the ground reveal little about them. But Io has always seemed an enigma. It is distinctly orange in colour, and since the 1950s, astronomers have known that it acts like a valve to control the strength of Jupiter's powerful radio broadcasts as the moon goes around the planet. Io lies within Jupiter's magnetosphere, and astronomers suspected that the moon must generate immense electric currents as it travels around Jupiter, like the metal coils in a power station generator.

Above: *A Roman mosaic from Vienne depicts the beautiful youth Ganymede being abducted by Jupiter, in the shape of an eagle. The astronomer Simon Marius gave the name Ganymede to the brightest and largest moon of the planet Jupiter.*

The volcanic moon Io hangs in front of Jupiter, in a spectacular picture from Voyager 2. Its eruptions have coated Io with orange sulphur deposits, which appear red in this colour-enhanced view.

The first space probes to Jupiter, however, had little chance to explore the moons. As their names suggest, Pioneer 10 and Pioneer 11 were trailblazers; their main purpose was to show that space probes could successfully make the long and dangerous journey to Jupiter.

The farthest that previous spacecraft had gone was to the planet Mars. The Pioneers had to travel eight times further, against the pull of the Sun's gravity. The improved Atlas Centaur rockets that launched the Pioneers in 1972 and 1973 blasted them away from Earth at a higher speed than any manmade object had previously travelled – sixteen times faster than a rifle bullet. This thrust was so powerful that it not only ensured a fast trip to Jupiter, but meant that Pioneers 10 and 11 would be the first objects to break free of the Sun's gravity altogether, and end up between the stars of our Galaxy.

Opposite: *In his book* The Starry Messenger, *Galileo recorded his pioneering observations of Jupiter's four largest moons, shown as small stars appearing to either side of Jupiter itself (large star).*

The first unknown that they faced was the asteroid belt. Between the orbits of Mars and Jupiter, there are multitudes of rocks of all sizes, the debris of a planet that never formed. The Pioneer scientists ensured that the craft avoided the known asteroids, but could there be a dense flock of small rock chunks that would damage the Pioneers? At their great speed, a collision with even a sand-grain could seriously damage the space probes. To the scientists' relief, first Pioneer 10 and then Pioneer 11 survived the seven-month passage through the asteroid belt, on course for Jupiter.

The Pioneers carried simple television cameras, which scanned Jupiter line by line as the spacecraft rotated, like the 'eyes' of the Viking landers on Mars. As Pioneer 10 swept past Jupiter in December 1973, at fifty-five times the speed of a rifle bullet, it sent back the first close-up pictures of a gas-giant world.

But the trailblazers' mission was as much to survive as to explore. Scientists wanted to send future missions – the Voyagers and Galileo – close in to Jupiter, where its strong magnetism traps a powerful torrent of electricity-charged particles in a 'radiation belt'. These particles could disable a spacecraft by damaging its sensitive electronics. As anxious scientists waited on Earth, Pioneer 10 swept past Jupiter's clouds – and lived to tell the tale. Pioneer 11 followed, swinging under Jupiter to take the first pictures of the planet's south pole, and it too survived the radiation belts.

The way was clear for bigger and more sophisticated spacecraft to explore Jupiter in detail: and the two Voyager probes were launched in 1977. That was a particularly favourable year for launching probes to the outer Solar System. The planets happened to be lined up in such a way that Jupiter's gravity could swing a space probe on towards Saturn; then Saturn could direct it to Uranus; and Uranus send it on to Neptune. American scientists originally proposed a special spacecraft to undertake this Grand Tour of the outer Solar System – but Congress shot it down as being too expensive. Out of the ashes rose the Voyager project, a pair of spacecraft based on the highly-successful Mariner missions to Mars and Venus.

Each of the Voyagers carried a television camera on a moving platform that could track its target as the craft shot through Jupiter's system, and would send back thousands of pictures. But this raised a problem. Radio signals take about forty-five minutes to reach the Earth from Jupiter; so a 'conversation' with a Voyager would contain gaps of one-and-a-half hours between sending a signal and getting a reply. With this kind of delay, it would be impossible to control the Voyager's television cameras from the Earth. As a result, all the camera angles had to be worked out beforehand, and the spacecraft themselves were equipped with powerful computers. These were the most sophisticated ever sent into space on an unmanned mission.

In the late summer of 1977, powerful Titan-Centaur rockets blasted the two Voyagers towards Jupiter. Voyager 1 completed its long journey in only eighteen months; while Voyager 2 (in fact, the first to be launched) took a slightly slower path and arrived four months later. Its more leisurely route would set it up for a journey that would eventually lead past Uranus and Neptune. As each Voyager approached Jupiter, it began taking a series of over 16 000 pictures of the planet and its moons.

The long sequence of pictures showed how the clouds of Jupiter were swirling round the planet – and it soon transpired that previous ideas were wrong. Jupiter has strong winds at particular latitudes that sweep the clouds along, in what seems at first to be a regular pattern, like the Earth's trade winds. But the Voyagers' cameras revealed that the motion of the clouds is much more turbulent. The wind-flows seem to arise from the constant turning of thousands of small 'eddies' in the planet's atmosphere. The Great Red Spot is a larger eddy, turning around every six days. Meteorologists now think that Jupiter's weather is controlled by the heat welling up from its interior as well as by the Sun's heat. The heat drives the eddies round and round, and they push the adjacent atmosphere along to create zones of wind, rather as the rotation of its wheels moves a car along the road. Our 'weathermen' know that the same thing happens in the Earth's atmosphere, but to a much lesser extent. The Voyager's picture of Jupiter's weather are thus helping us with the very down-to-earth problem of understanding the Earth's weather patterns.

Meteorologists expected to learn a lot about Jupiter's weather from the Voyagers; but no one expected to find a ring around the planet. Toby Owen, one of the scientists studying the clouds of Jupiter, thought that it might be an idea to take a picture of the

A mosaic of pictures from Voyager 2 shows Jupiter's narrow ring (yellow) stretching out to either side of the planet. Voyager took this photograph above Jupiter's night-side, so the planet shows only as a circle of red light where its atmosphere is illuminated by sunlight. (Parts of this circle, and of the ring, are missing, in regions where pictures are missing from the mosaic.) The small white specks are intense flashes of lightning in Jupiter's atmosphere.

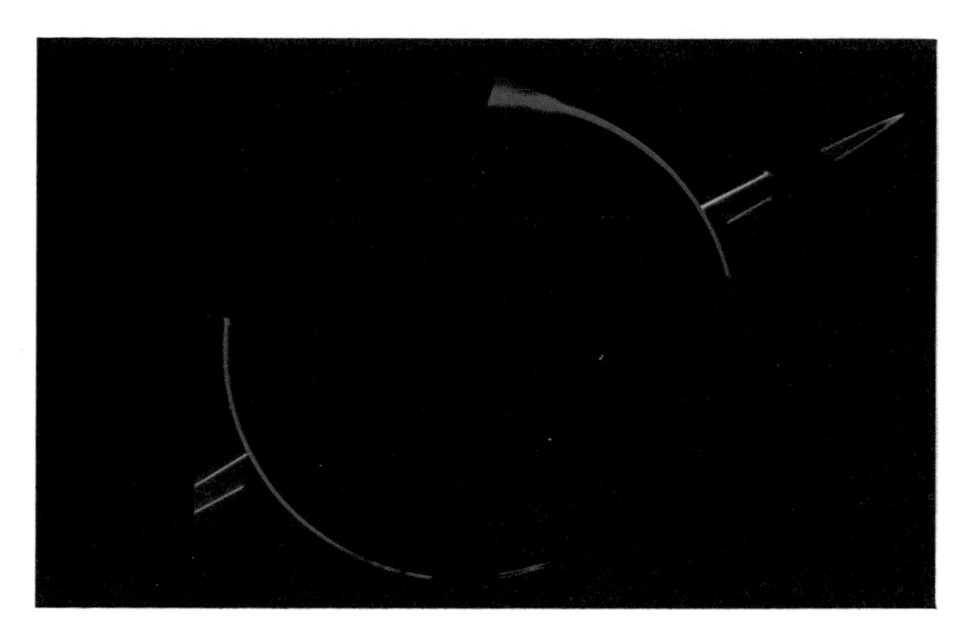

empty space outside the planet's equator – just in case. Computer programmer Candy Hansen ensured that Voyager 1 had this command in its brain before it reached the best viewpoint for a possible ring. As that picture came back and built up on the television screen, Hansen saw a streak across the screen: 'Hey, Toby, we got something,' she called. To Owen's surprise as well, they had actually found a very faint ring girdling Jupiter's equator.

Voyager 1 did not have a second chance to view the ring, but the discovery gave the Voyager team plenty of time to instruct the second craft to take pictures of it. Voyager 2 found that the ring has a sharp outer edge, and stretches down, fainter and fainter, towards the planet's cloud tops. When the probe viewed the ring from behind Jupiter, looking at what should be the dark side, the ring actually appeared brighter. This contradictory behaviour is typical of very small particles, like the smoke from a cigarette. They must be spiralling down towards Jupiter from the ring's outer edge. But where do they come from?

The mystery of Jupiter's ring began to clear up when other Voyager scientists discovered that the cameras had unwittingly photographed three new moons. These small satellites form a quartet, with Amalthea, of moons orbiting much closer to the planet than the great Galilean satellites. And the two innermost moons lie right at the edge of the rings. These moons – and perhaps smaller moons in the region of the rings – are being bombarded by tiny meteorites falling towards Jupiter. The impacts can chip off small fragments of rock, which spread out to make Jupiter's ring.

One of the Voyager's main tasks was to investigate the four Galilean satellites. Astronomers had every reason to think that these four similar-sized worlds would be very similar to one another. Nothing could be further from the truth. Among these four bodies, the Voyagers found some of the most extreme surfaces in the Solar System: the smoothest, and the most-cratered; the oldest, and the youngest.

The outermost of the four, Callisto, is a world of frozen slush. It is made of a mixture of rocks and ice, and its surface looks correspondingly dirty. In the early days of the Solar System, infalling meteorites pocked the surface of Callisto with crater upon crater. On other worlds in the Solar System craters have been erased by lava flows or by other mechanisms. But Callisto has stayed dormant since its birth. The jumble of craters, covering every inch of the surface, show that this moon has the oldest unaltered surface in the Solar System. Callisto is a fossil from the Solar System's birth.

Ganymede, the largest of the Galileans, is a piebald moon. It has large dark areas, heavily cratered, which look like Callisto – and are probably remnants of Ganymede's original surface. Like the Earth's continents, these regions have been split apart from one another. The gaps between the dark 'continents' consist of ridges and valleys, sculpted from lighter material, probably cleaner 'slush'. Geologists and planetary scientists still don't understand what's happened on Ganymede. The breaking of its crust is rather like continental drift (plate tectonics) on the Earth, but the actual flow

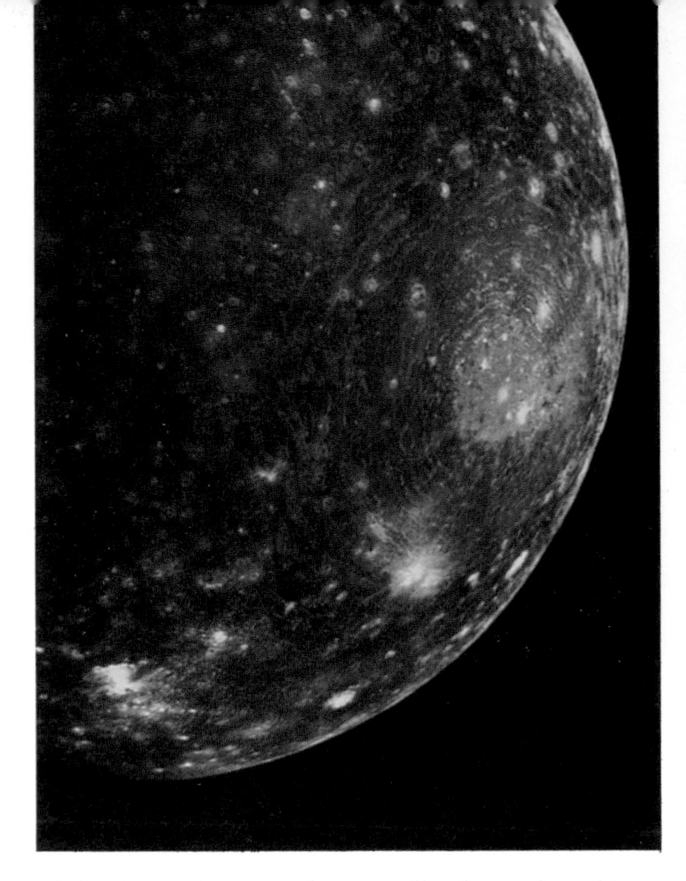

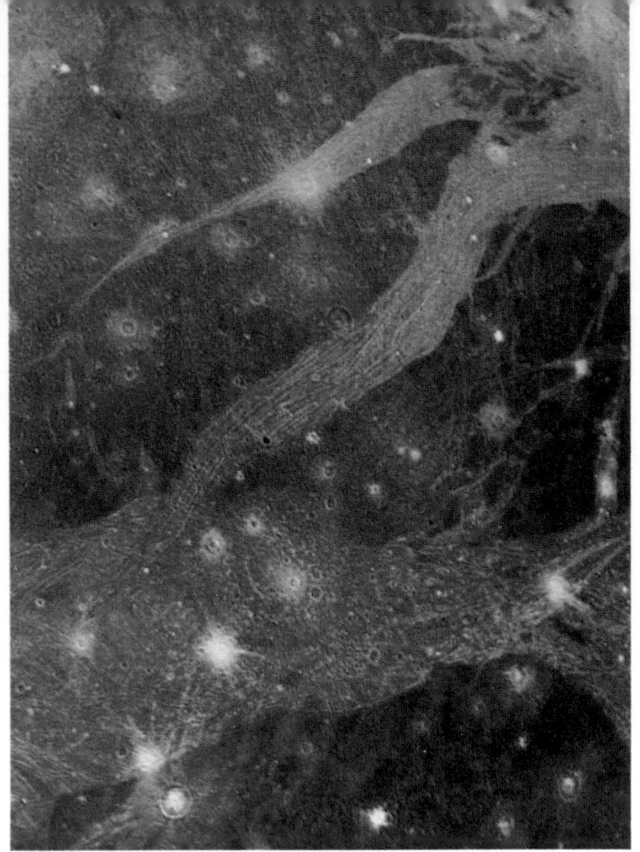

of the frozen slush must be more like the motion of large glaciers and the great ice sheets that cover Greenland and Antarctica.

The appearance of Europa came as a surprise for the Voyager scientists. About the size of our Moon, it is a bright little world, reflecting sunlight as dazzlingly as an icy pond on Earth. And the Voyager images show that Europa's surface is just that: a smooth sheet of ice. For some unknown reason, Europa has heated up inside. As the ice within melted, it welled up to cover the entire surface with oceans – and in the extremely cold conditions so far from the Sun, the top of the oceans froze to form a smooth icy surface. Like pack-ice on the Earth, Europa's ice-sheets have cracked in places; dirty water has welled up and frozen in between the great floes to produce a delicate lacework pattern across the moon. All the original craters are hidden under its frozen sea, and despite the cracks in the ice, Europa's surface has ended up as smooth – compared to its size – as a billiard ball.

If Europa was a surprise, then Io was a shock. Astronomers knew that Io was yellowish-orange in colour, and this coloured material was probably sulphur. But this had not prepared them for the sight that Voyager showed: enormous volcanoes erupting all around Io, spewing sulphurous vapours 200 miles (300km) upwards. For its size, Io is the most active body in the Solar System – and it vies with Venus for matching the traditional 'fire-and-brimstone' vision of Hell.

Io is almost identical in size to our Moon, and by rights it should be a cold, dead world. Its central fires are raised by the effects of Jupiter's gravity, and that of the neighbouring moon, Europa. As Io goes around Jupiter, the giant planet's gravity distorts it into an egg-shape, always pointing more-or-less toward Jupiter. If Io's orbit were a circle, nothing special would happen; but Io follows an elliptical (oval) path around Jupiter. Io is most stretched when it is closest to Jupiter, and less elongated at the far point of its orbit. Because its speed also changes along its orbit, Io's bulge at times turns slightly away from the direction of Jupiter; and the planet's gravity then twists Io, by pulling on the bulge.

The poor moon is continuously kneaded by these forces, and it would normally duck out of the situation by making its orbit more circular. But that doesn't allow for Europa. It goes around Jupiter at exactly twice the period of Io's orbit. So Io and Europa keep passing at their closest in a regular fashion, at every orbit of Europa and every other orbit of Io. Europa's gravity provides a small pull that prevents Io's orbit from becoming circular.

Left: *The dark surface of Jupiter's moon Callisto is completely covered in craters. An exceptionally powerful impact has fractured its icy surface, and created an enormous 'bull's-eye' of cracks, called Valhalla.*

Above: *Ganymede's dark surface is interrupted by lighter bands, consisting of parallel grooves and ridges. They are as large as some of the Earth's mountain ranges, and probably result from the motion of parts of Ganymede's crust, similar to 'continental drift' on the Earth.*

Europa is a very smooth and bright world. This Voyager 2 picture has been contrast-enhanced to show the pattern of cracks that cross its icy surface. There are no craters visible, meaning that Europa's original cratered surface must have melted and then re-frozen.

The constant kneading of Io heats up its centre, just as you can make a metal wire hot by bending it back and forth. For its size, Io produces more heat than anything in the Solar System, bar the Sun! A few weeks before Voyager 1 reached Jupiter, a team of American astronomers calculated that this effect could melt rocks within Io – but even they did not predict the startling volcanic landscape that Voyager revealed.

Voyager 1 found eight enormous volcanoes erupting on Io. Their gases rose like enormous umbrella-shaped plumes, spraying the surface with sulphur and sulphurous compounds in shades of orange, yellow and white; lakes of molten and solidifying lava spot the world with smaller dark spots. The overall effect is to make Io appear like a cosmic pizza.

The volcanic moon Io leaves a trail of gases in its wake as it orbits Jupiter. This picture combines a real photograph of the glow from sodium atoms near Io (glowing yellow, like sodium streetlamps) with a computer-generated outline of Jupiter and Io's orbit.

Sue Kieffer of the US Geological Survey points out that the name 'volcano' is perhaps misleading – the eruptions on Io are actually geysers, giant versions of the famous Old Faithful in Yellowstone National Park. In Old Faithful, a deeply buried pocket of molten rock heats up an underground pool of water, and the water erupts upwards when it boils. The molten rock doesn't erupt itself, but passes its heat on to the water, which boils more easily. Similarly, on Io, the hot molten rock in its interior does not make its way to the surface. It heats up hidden pockets of sulphur, which can then erupt as great volcanoes like Pele, named for the Hawaiian fire goddess. But Io is a

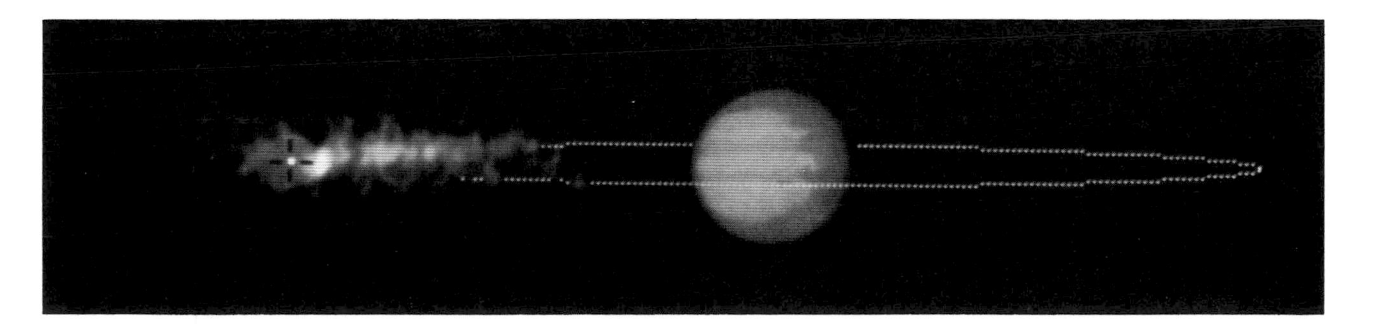

more complicated place than the Earth. The molten sulphur may not reach the surface itself. As it rises, it could meet a pocket of frozen sulphur dioxide, and heat this substance so that a sulphur dioxide geyser breaks out, to form one of Io's smaller eruptions.

Kieffer isn't worried that Io's eruptions are so much larger than any geyser on Earth. She points out that Io has a gravity only one-sixth as strong as the Earth's, so any eruption would go six times higher; and that gases from a geyser on Io are erupting into a vacuum, where they will spread out much more than an eruption into Earth's atmosphere. If we were to take Old Faithful to Io, it would not be 100 feet (30 metres) high, but twenty miles (30km)!

Io's tortured volcanic surface is revealed in this dramatic, true-colour photograph from Voyager 1. Occasional sharp-sided hills protrude from the sulphur-yellow surface, spotted with volcanic calderas and dark lava lakes. The bright complex of hills at upper right is Io's largest volcano, Pele. Its plume of gases is invisible against Io's bright surface, but appears at the extreme upper right against the dark sky – rising 200 miles (300km) above the volcano.

The Voyagers' discoveries at Jupiter have only whetted astronomers appetites to find out more about this 'Solar System in miniature'. The next American expedition is now at an advanced stage of preparation; and it should reach Jupiter in December 1988.

Fittingly enough, this new spacecraft is called Galileo. While previous space probes have flown past Jupiter in a matter of days, Galileo will go into orbit, as an artificial moon of Jupiter. For at least two years, the probe will watch the constantly-changing weather patterns on Jupiter, and monitor the eruptions on Io.

The space shuttle will – according to present schedules – take Galileo up into Earth orbit in May 1986. A powerful Centaur rocket will then boost Galileo on its two-and-a-half year flight to Jupiter. Galileo will travel more slowly than the Pioneers and the Voyagers for a very good reason. When it arrives, it will despatch a smaller probe into Jupiter's atmosphere – and a quicker journey would give that probe a faster passage into Jupiter's atmosphere, raising the danger that friction will incinerate it like a meteor.

Five months before reaching Jupiter, when Galileo is targeted straight for the planet, it releases the spinning atmosphere probe. Like a bullet, the probe heads for Jupiter, while the main spacecraft changes its course to miss the planet by a small margin. Pulled by Jupiter's gravity, the probe speeds up, until it hits the top of the atmosphere

An artist's impression shows the Galileo spacecraft dropping off its probe to explore Jupiter's atmosphere (right). The main craft, with its big umbrella-like antenna to communicate with Earth and its nuclear power-generator on the upper boom, will send back pictures of Jupiter and its moons for two years.

at a speed of 115 000mph (185 000kph). Its front is heated to incandescence, like a meteor hitting the Earth: if there were anyone in Jupiter's clouds to watch, they would see the probe shine as brightly as the Sun. The atmosphere in front of the probe is heated to 16 000°C – three times hotter than the Sun's surface – but a specially designed heatshield should protect the rest of the probe. The impact slows the probe so abruptly that the 'G force' on it is 350 times stronger than Earth's gravity.

The whole ordeal lasts only two minutes. Then the remains of the heatshield, half-burnt away, are dropped and the probe parachutes gently through the successive layers of clouds. Unfortunately, it cannot carry a camera to record the spectacular views in Jupiter's clouds, but the probe will tell us exactly what those clouds, and Jupiter's atmosphere, are made of; and it will measure temperatures and pressures to add to our weather map of Jupiter.

Meanwhile, the main Galileo spacecraft will swing into orbit around Jupiter. It cannot carry enough fuel to slow down its flight past Jupiter. Instead, Galileo will use Io's gravity to kill its forward momentum, swinging only 600 miles (1000km) above the moon – just three times the height of the largest volcanic plumes. During the next two years, Galileo will pass closely by the other Galilean moons, using their gravity to keep changing its path, into a complex petal-shape of orbits.

As well as watching Io's volcanoes, Galileo's cameras will carefully study the other Galilean moons. They will show objects as small as a city block on all four moons, giving us views of these distant worlds that are as detailed as the Landsat pictures of the Earth. Galileo can also show images in infra-red radiation. This instrument will show directly the hot and cold parts of volcanic Io, and indirectly it can reveal the composition of the rocks and ice of the moons' surfaces.

Geologists will be able to work out just what has happened on cratered Callisto, and the meaning of the stretch-marks on Ganymede. Vulcanologists will watch how the great eruptions on Io grow and die down, turning this little moon inside out.

But most exciting may be its views of the ice-covered world Europa. The project scientist for Galileo, Torrence Johnson, says, 'Europa will do for Galileo what Io did for Voyager.' We know that this billiard ball of a world is covered in ice, but we don't know how thick it is. Just as a winter's frost can freeze the surface of a lake on the Earth, but leave plenty of liquid water beneath, so Europa's crust may hide oceans of water. In that case, it would be the only other known world in the Solar System with liquid water. And if the Earth's oceans were able to bring forth living creatures on our planet, might there be a remote chance that life of some kind has begun in the chilly seas beneath Europa's icy surface?

At the moment, this is only speculation – a speculation brilliantly portrayed in the film *2010* – but the Galileo mission will tell us whether such ideas are merely science fiction, or science fact.

CHAPTER EIGHT

SATURN

Ask any astronomer which is the most beautiful of the planets, and the answer would undoubtedly be 'Saturn'. Even a small telescope reveals that Saturn is unique: a golden globe encircled by an exquisite set of lustrous rings. Its symmetry and perfect proportions make it appear like a small three-dimensional model suspended at the top of the telescope tube. The close-up portraits of Saturn snatched by the Voyager spacecraft only enhance the planet's reputation. The rings are finely etched with thousands of dark lines, as closely-spaced as the grooves on a long-playing record; and they are as flimsy as fine lace, for in some photographs the Voyagers' cameras show us the planet's globe shining eerily through the rings.

A Roman statue, dating from the first century AD, depicts the god Saturn with a cornucopia – a 'horn of plenty'.

Saturn's beauty stands in stark contrast to its namesake god and his astrological associations. He was the father of Jupiter, and one of the Titans, the elder generation of gods. The Greeks and Romans portrayed Saturn (or his Greek equivalent Kronos) as an elderly figure, which in later times became Old Father Time, and it seemed natural to associate him with the slowest-moving planet. Like many of the ancient gods, Saturn married his sister – the Titaness Rhea – and showed his unpleasant character by swallowing his children, to avert a prophecy that his own son would depose him. Rhea managed to save Jupiter, however, and he did indeed take Saturn's place as king of the gods. Medieval astrologers endowed the slow-moving planet with appropriately sluggish and malevolent attributes, summed up in the adjective 'saturnine'. For its metal, they chose dull and heavy lead.

The old ideas were wrong about more than just Saturn's appearance. The 'sluggish' planet has hurricane-force winds that constantly whip around its equator; and far from being lead-heavy, Saturn has the lowest density of all the planets – so insubstantial that it would float on water, if we could find an ocean big enough.

Far from being saturnine, the planet is a tease. Unlike Venus, which simply hides its secrets beneath a cloudy veil, Saturn reveals just enough for us to think we understand. Then when we get a better view, we find that our previous ideas were wrong – and the planet throws up another surprise which seems to cast doubt not only on our theories of Saturn but also apparently on science itself. Saturn caught out Galileo in 1612, when he looked for two moons he had seen clearly a couple of years earlier, and found no trace of them: 'I do not know what to say in a case so surprising, so unlooked for, and so novel,' he mused. Nearly 370 years on, Brad Smith, one of the Voyager scientists, echoed Galileo's sentiments. The spaceprobes sent back a picture that

showed a triplet of narrow rings, seemingly braided around one another; and Smith confessed, 'It boggles the mind: the kinks and braiding seem to defy the laws of orbital motion.'

One mystery of Saturn reverberates down to us from the beginning of our era. Was it responsible for the Star of Bethlehem? For centuries, historians and astronomers have debated what kind of an object the 'Wise Men' saw in the sky, a 'star' so unusual that it signified the birth of our Saviour. Scholars agree that Christ was born a few years BC (because a medieval monk made a mistake in calculating our BC/AD dating system). Astronomers of that time recorded only two unusual celestial sights that could have inspired the Magi. In 5 BC, an exploding star, or nova, appeared in the constellation Capricorn and shone for seventy days. Although this was a rare event, which was recorded in detail by Chinese astronomers, it probably meant little to the Wise Men, whose astrology was based on the planets.

And indeed, in 7 BC there was an unusual planetary sight: a triple conjunction of Saturn and Jupiter. Saturn, the most distant planet known to the ancients, takes 29½ years to complete one orbit of the Sun. As a result, its motion around the sky appears as a tortoise-crawl from constellation to constellation, in which almost three decades elapse before Saturn returns to its starting point. This slow pace is temporarily reversed for a couple of months each year, and the planet seems to make a loop in the sky as the fast-moving Earth overtakes Saturn. Jupiter's motion follows a similar pattern, but at a faster rate, so that it completes one circuit in only twelve years. The faster-moving Jupiter catches up with Saturn every twenty years. Every so often, one of these conjunctions has a more interesting character: if the Earth, Jupiter and Saturn are almost in line, the apparent loop of Jupiter will be superimposed on Saturn's looping path. From our viewpoint on Earth, the two planets seem to approach closely three times. Such triple conjunctions are relatively rare, occurring only every 139 years.

To the astrologers of the Middle East, the planet Jupiter represented the king of the gods, and Saturn was the planet that protected the Jews; moreover, the triple conjunction of 7 BC occurred in Pisces, a constellation that was also associated with the Jews in astrology. Seeing this portent in the sky, the Magi naturally came to Jerusalem and asked of Herod 'where is he that is born King of the Jews'?

The idea is appealing – yet we cannot be certain. The two planets never came closer together than two Moon-widths in the sky, so they never appeared to amalgamate as a single, more brilliant star. And St Matthew's gospel consistently speaks of 'the star', not 'the stars' or 'the planets'. There's also a chance that the star never existed. St Matthew may have included one in his account so that the gospel would fulfil an Old Testament prophecy.

As it approached Saturn in July 1981, Voyager 2 showed details of the ringed planet that are never visible from the Earth. This true-colour photograph shows the planet's dull yellow colour, which makes it shine balefully in our skies.

The scientific study of Saturn began when Galileo turned his telescope to the planet in 1610. He had already found the four moons going around Jupiter, and he noticed a large 'moon' on either side of Saturn. But to his surprise, the two moons did not move from night to night, and they were fully one-third as large as Saturn itself. The practical Galileo wrote, 'I have observed the furthest planet to be triple' – and the discovery was so odd that he published the sentence as an anagram. This meant he could claim priority if anyone else discovered Saturn's triple nature; yet if he was wrong, no one need ever know!

Two years later, Galileo was even more puzzled. His telescope showed him just a single world: Saturn's two companions had disappeared. Recalling the mythology, he wondered, 'has Saturn, perhaps, devoured his own children?' When he looked at Saturn again, in 1616, he found the companions were back, but now shaped as 'two half-ellipses'. With hindsight, a modern astronomer could easily put Galileo out of his confusion. The appearance of the rings changes as Saturn goes round its orbit. Galileo first saw the rings at a shallow angle, so he could make out the two extremities but not the foreshortened portions running behind and in front of Saturn. In 1612, the thin rings were edge-on to us, and so were invisible. By 1616, they had opened out again, and Galileo gave an almost correct description.

Previous page: *Computer-enhancement brings out the very subtle shades of colour in Saturn's rings, as seen by the Voyager spacecraft, to create a gaudy – but scientifically useful – image. The Voyagers showed that the rings are divided into thousands of narrow ringlets.*

Over the next four decades, other astronomers drew Saturn's changing appearance, and some of these were remarkably accurate. But without the idea of a ring in mind, no one could make the crucial step in interpreting these observations. The favourite idea involved Saturn exhaling vapours that condensed around the planet – rather like the volcanic plumes that the Voyager spacecraft were to discover much later on Jupiter's moon Io.

The leap in imagination came indirectly. The Dutch astronomer Christiaan Huygens invented a new way of making telescope lenses, and his instruments gave much sharper views of the planets. In March 1655, he found that Saturn does have a genuine moon, Titan. As Huygens followed Titan around Saturn, he discovered that its orbit was tilted, compared to Saturn's orbit around the Sun. Assuming that Titan went around Saturn's equator, that meant the planet itself was tilted. So perhaps its changing appearance was not due to alterations in Saturn itself, but to our different angle of view as Saturn goes around the Sun. He soon realized that all the drawings could be explained if Saturn were 'surrounded by a thin flat ring, nowhere touching, and inclined to the ecliptic'. (The ecliptic is the plane of the Earth's orbit about the Sun, which is almost identical to Saturn's.) Like Galileo, he first published the sentence as an anagram; by 1659 he was confident enough to publish it unencoded.

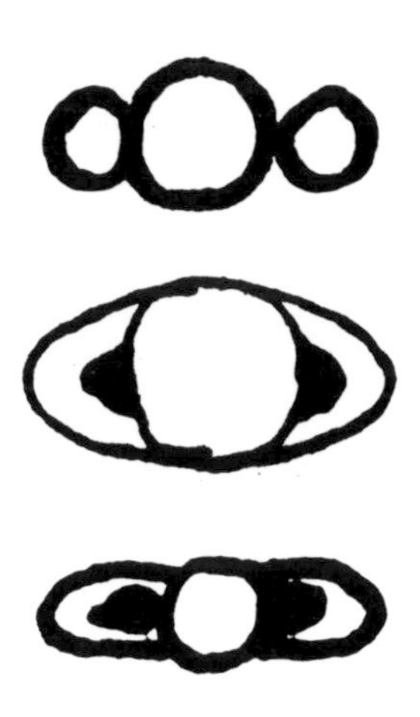

With his primitive telescopes, Galileo could not make out the true nature of Saturn – though he came tantalisingly close. These three drawings are from a letter he wrote to his friend Marcus Welser in 1612.

Helped by Huygen's explanation as much as by better telescopes, everyone could see the ring. In 1675, Giovanni Cassini – who ten years earlier had seen Jupiter's Great Red Spot – found that the ring was double: a dark line ran round, separating an outer ring from an inner. The gap between the rings is now called Cassini's Division. In 1850, several astronomers could see the planet through the ring.

Saturn had set another puzzle: what are the rings made of? At first astronomers thought they were simply solid sheets, like the brim of a top hat. But as physicists came to understand more about gravity, they realized that such a sheet would have problems in orbiting Saturn as a whole. In 1857, the great British physicist James Clerk Maxwell proved that the rings cannot consist of solid sheets, or sheets of liquid: they must consist of innumerable separate particles, orbiting Saturn as miniature moons. All our discoveries and observations since then have reinforced Maxwell's deduction, yet even the Voyager probes were unable to detect the individual particles in Saturn's rings. It's a reminder of the power of the human brain that a physicist working well over a century ago could tell us some facts about a distant world that even our most sophisticated space probes cannot reveal directly.

In the meantime, improving telescopes were allowing astronomers to pick up more satellites in orbit about the ringed planet. Cassini discovered four, including the strange moon Iapetus which is much brighter when on one side of Saturn than on the other. William Herschel, the discoverer of Uranus, found two more moons of Saturn, and another, Hyperion, was discovered simultaneously by George Bond at Harvard and the amateur astronomer William Lassell in England. Another Harvard astronomer, William Pickering, increased Saturn's family to nine, when he photographed faint Phoebe in 1898.

The great Dutch scientist Christiaan Huygens was the first to realize that Saturn is surrounded by a narrow ring. Here he is shown with another of his contributions to science – the first pendulum clock.

Other astronomers were trying to find out about Saturn's larger moons, especially giant Titan, which is bigger than the rest of Saturn's moons put together. The harder they looked, the stranger Titan seemed. In 1944, Gerard P. Kuiper at Chicago University found that Titan has an atmosphere. Larry Trafton used the telescopes of the McDonald Observatory in Texas to follow up this discovery, and found that Titan's atmosphere must be surprisingly dense, and must be pretty hazy. Although we now know that Jupiter's moon Io and Neptune's Triton have some atmosphere, Titan is unique in having such a dense blanket. More theoretical work showed that Titan should have clouds. When American scientists began to plan space probes to fly past Saturn, many of them were as keen to explore Titan as they were to investigate Saturn and its rings.

When the space probes left, Saturn was even more of an unknown than Jupiter. To be sure, astronomers knew the basic facts about Saturn. It is a giant world, heavier than all the other planets except Jupiter put together. Yet its ninety-five Earth-masses of material comprises a relatively large globe, so its density is correspondingly low – lower than the density of water. Like Jupiter, Saturn is a world made of gases, mainly hydrogen and helium. It spins rapidly, with a day of only ten hours forty minutes, and the rapid rotation distends Saturn's lightweight gases into a globe even more flattened than Jupiter's. But that was about the limit of our knowledge. Saturn presents a bland face to the Universe. It has none of the bright colours and gaudy whirls and spots that Jupiter displays. Saturn's globe has only a few narrow dark belts running parallel to its equator. Occasionally, a white spot will flare up and be carried around by the planet's rotation. Keen amateur astronomers have sometimes spied these spots before the professional astronomers; a great white spot that erupted in 1933 was caught by, amongst others, Will Hay, the British film actor.

The small Pioneer 11 craft could look forward to a host of new discoveries at Saturn, as well as scouting out the territory for the two Voyagers to follow. The mission controllers aimed its approach to Jupiter in December 1973 in such a way that the giant planet's gravity yanked the small space probe around, and sent it hurtling back across the Solar System, towards a rendezvous with Saturn. This graphically-named 'gravitational slingshot' sent Pioneer 11 looping upwards, out of the plane of the Solar System. It flew high over the asteroids before its momentum carried it outwards again, across Jupiter's orbit and on to Saturn. When it reached Saturn, Pioneer 11 was six-and-a-half years old, the survivor of the longest interplanetary trek up to that time. NASA's Ames Research Center, which controlled the Pioneers, had to recall six key members of the staff who had retired during the five years since Pioneer 11 passed Jupiter!

As Pioneer 11 swung past Saturn in August and September 1979, its simple cameras revealed new aspects of the rings – quite literally, for it scanned the side of the rings that is not lit by the Sun. Pioneer found that Cassini's Division is not entirely empty: it contains a thin sprinkling of particles and appears dark only by comparison with the bright neighbouring rings, the outer A-ring and the brilliant B-ring within the division. Seen on its own, Cassini's Division would shine faintly in reflected sunlight, like the Crepe ring, now known less romantically as the C-ring. Its camera also showed Earth-based astronomers were wrong in suspecting a fainter 'D-ring' within the C-ring.

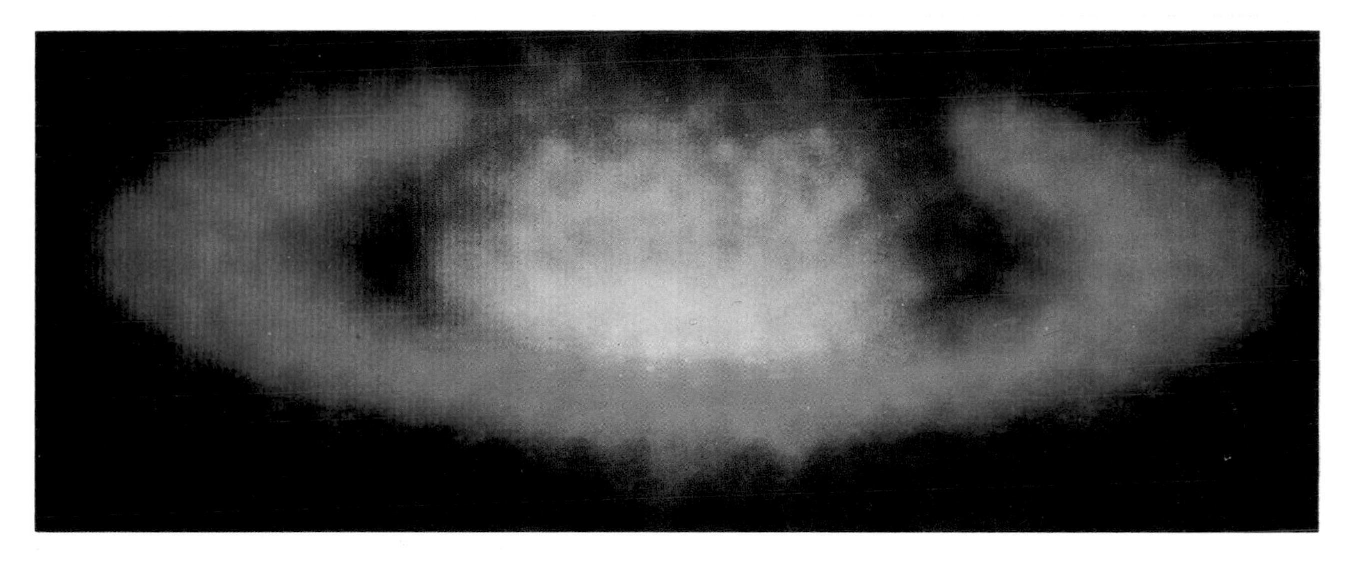

A false-colour 'heat image' of Saturn is coded so that it shows the Sun's light and heat reflected from the rings in blue, while the heat welling up from within Saturn makes the planet's globe glow in yellow. Saturn's output of heat is probably due to a 'rain' of helium at its core.

But the Pioneer scientists were more interested in discovering new rings. And they were not disappointed. As Larry Esposito watched the television pictures build up line by line he saw a bright blip outside the main outer A-ring; successive scans showed it reappear, and build up into a curving line that followed around the edge of the A-ring, some 2500 miles (4000km) further away from the planet. It was something completely unexpected, a narrow ring encircling Saturn. Hailed as a surprise at the time, the F-ring was to gain notoriety when Voyager 1 took a close look at it.

Pioneer's most exciting discoveries about Saturn came from other instruments on board. Its magnetometer showed that Saturn has a powerful magnetic field, a thousand times stronger than the Earth's. The discovery pleased scientists like Bill Hubbard, of the University of Arizona, who had calculated that Saturn's interior should resemble the inside of Jupiter. Apart from a small rocky core, Saturn is a liquid planet, consisting mainly of hydrogen and helium. The inner parts of the planet's deep 'ocean', nearest to the core, are compressed so much that the hydrogen behaves like a metal. Immense electric currents flow through this liquid metal, generating the magnetic field.

A navigator on Saturn would find its magnetism particularly useful. All the other planets with magnetic fields have their magnetic poles some distance from their true poles – the Earth's magnetic poles, for example, lie in northern Canada and at the edge of Antarctica – but Saturn's magnetism is precisely lined up with the poles of rotation. We would not need to make corrections to a magnetic bearing taken on Saturn, because the compass needle points due north-south. Convenient though this would be for a hypothetical Saturn traveller, it leaves planetary scientists rather perturbed. Their detailed theories on how a planet's liquid metal interior can produce a magnetic field all predict a magnetic field that is skewed to the planet's rotation. Saturn's due north-south magnetism has set another conundrum for astronomers.

Pioneer 11 also measured the amount of heat that Saturn radiates into space, and uncovered a further puzzle. Saturn is producing heat in its interior. At first sight, that does not seem surprising, because astronomers already knew that Saturn's larger counterpart, Jupiter, is generating heat as it gradually shrinks to its final size. But more detailed calculations showed that the smaller planet Saturn should have reached its final size over two billion years ago.

Was this excess heat perhaps just a misreading by the ageing instruments on Pioneer 11? The later Voyager probes confirmed that Saturn does indeed emit more heat than it receives from the Sun, though not quite as much as Pioneer had shown. The amount was reduced still further when astronomers recalculated how much of the Sun's heat Saturn reflects back into space. But Bill Hubbard admits 'there's still an excess, though not as dramatic'.

Flight controllers at NASA's Jet Propulsion Laboratory in California are in constant contact with all American probes to the planets. Here they are receiving Voyager 1's pictures of Saturn, sent back from almost 1000 million miles (1400 million km) away.

Dave Stevenson of Caltech in Pasadena is an expert at calculating how matter behaves in the crushed interiors of planets. He has calculated that something strange must have happened in Saturn's core, once it had stopped shrinking and had begun to cool down. The planet's liquid metal interior was originally a mixture of hydrogen with a smaller amount of helium. But when the temperature dropped, the helium became insoluble in the liquid hydrogen metal, and it began to separate out. Being heavier than hydrogen, the drops of helium began to fall towards the planet's centre, as a helium 'rain' within the hydrogen ocean. You can see the same thing happening, in a more mundane way, when you shake a bottle of salad dressing and let it stand: the heavy drops of vinegar (helium) fall through the lighter oil (hydrogen) and collect at the bottom of the bottle (around Saturn's rocky core). What isn't immediately obvious in the kitchen is that the fall of the drops generates some heat. On a small scale it may be unnoticeable; but when we rearrange the contents of a planet of Saturn's size, the heat is tremendous.

By the time the two Voyager craft were launched, the outer planets had rolled around into more convenient positions. Although it set off from Earth more than four years after Pioneer 11, Voyager 1 reached Saturn little more than a year after the Pioneer had passed by. Its sophisticated television cameras were instructed to lay bare the secrets of Saturn, its rings and its moons – especially Titan, the moon with an atmosphere.

False-colour versions of the Voyagers' photographs show up the weather patterns of Saturn. The largest of the three spots is a high pressure region, of the kind that brings fine weather in the Earth's atmosphere. The jagged border of the yellow cloud layer is a 'jet-stream' of high-speed winds.

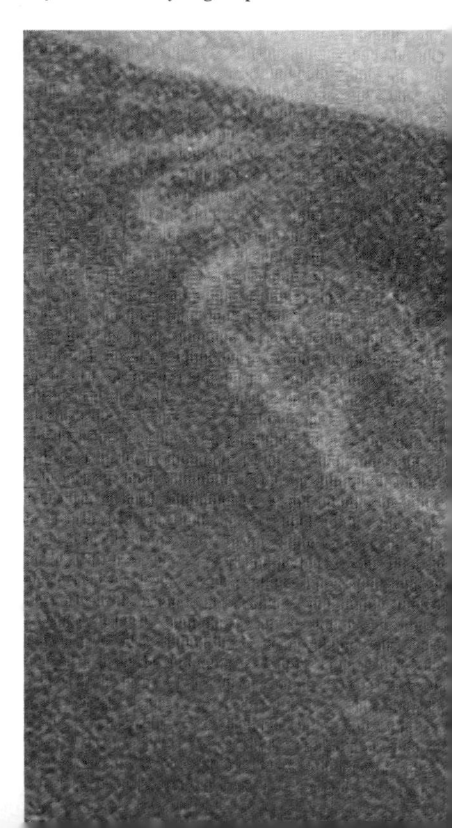

As Voyager 1 closed with the planet, scientists on Earth suffered their first disappointment. Saturn doesn't have the brightly-coloured spots, eddies and bands that Jupiter displays. Voyager scientist Anne Bunker found one red spot, but 'Anne's Spot' was the exception. Saturn seems to be covered with a continuous layer of frozen ammonia cirrus that hides any cloud patterns below. Its golden-orange colour looks as though someone has taken Jupiter's bright oranges and browns, and mixed them with its pure white zones, to create a uniform layer of colour over the planet. That may be what has happened in some way that is, as yet, only vaguely understood.

Unlike our eyes, the Voyager's cameras could see more than just visible light. They were sensitive to ultraviolet light, too. The frustrated Voyager scientists combined the ultraviolet and visible pictures to create false-colour images that enhanced the tantalizing details in Saturn's clouds. A sequence of these pictures enabled meteorologists to piece together the general patterns of Saturn's weather systems.

They first confirmed a fact that Earth-based astronomers had suspected for many years. The clouds at Saturn's equator race around the planet at a tremendous rate, swept by the fastest winds in the solar system – constant 1100mph (1800kph) winds that blow four times faster than Jupiter's equatorial winds and ten times more rapidly than the fastest winds we ever experience on Earth. Away from the equator, north and south, Voyager 1 found that Saturn has alternating bands of slow and fast winds. To the meteorologists' surprise, these winds follow exactly the same pattern in each

hemisphere. Andrew Ingersoll has suggested that the bands of similar speed winds in each hemisphere are actually connected right through the planet. So Saturn's weather is not confined just to the thin upper layer where the clouds lie, but extends deep into the immense liquid oceans that make up most of the planet.

If Voyager 1 could prise few secrets out of Saturn, it had even less success with Titan. Following instructions, the cameras took picture after close-up picture of Titan. Scientists hoped they would show the suspected wisps of cloud, and details of Titan's surface beneath. In fact, the pictures showed nothing but clouds, and more clouds: a continuous layer of orange cloud that covers the whole world.

The cameras were only part of the Voyager's armoury. Other instruments could give some clues to the world beneath the clouds. For decades, astronomers had believed Titan to be the largest moon in the solar system: the Voyager's results, however, relegated it to second place, just behind Jupiter's satellite Ganymede – but even so, Titan is still a little larger than the planet Mercury. As Earth-based astronomers had suspected, Titan's atmosphere is surprisingly thick for so small a world: at the surface, its pressure is almost twice that of the Earth's atmosphere.

There was another surprise. Titan's atmosphere consists mainly of nitrogen – the substance that makes up four-fifths of the Earth's atmosphere. Our near neighbours in space, Venus and Mars, have predominantly carbon dioxide atmospheres; the four giant planets have hydrogen-helium atmospheres. The only twin of the Earth's atmosphere is to be found in this tiny frozen world circling distant Saturn.

The extreme cold at Saturn's distance from the Sun makes Titan a rather different world from the Earth. For a start, water is frozen into ice, and Titan's body is probably a half-and-half mixture of ice and rock. But if water is not liquid on Titan, another substance may take its place. Mixed in with the nitrogen in its atmosphere is a small amount of methane gas. At Titan's temperature of minus 180°C, methane is likely to condense out of the atmosphere, and fall as 'rain' or 'snow'. The landscapes of Titan could be surprisingly similar to the Earth's, with the planet's rocks replaced by an ice-rock mixture, and water by methane. Methane rain drips from clouds (hidden well below the dense orange clouds), and collects into methane seas and oceans. Near Titan's poles, methane snow falls on to vast icecaps of frozen methane, which calve 'methanebergs' into the adjacent ocean.

Voyager scientist Brian Toon envisages sitting in a boat on one of Titan's methane seas. It's much darker than we are used to on Earth, partly because Titan is further from the Sun, and partly because of the clouds. The sunlight filtering through is about as bright as a moonlit night on Earth, and shows us the eerie shape of the bergs floating by. What fuel should we use to power our boat? There's no problem for we are floating on an ocean of fuel – methane is the main ingredient of the natural gas that we burn on

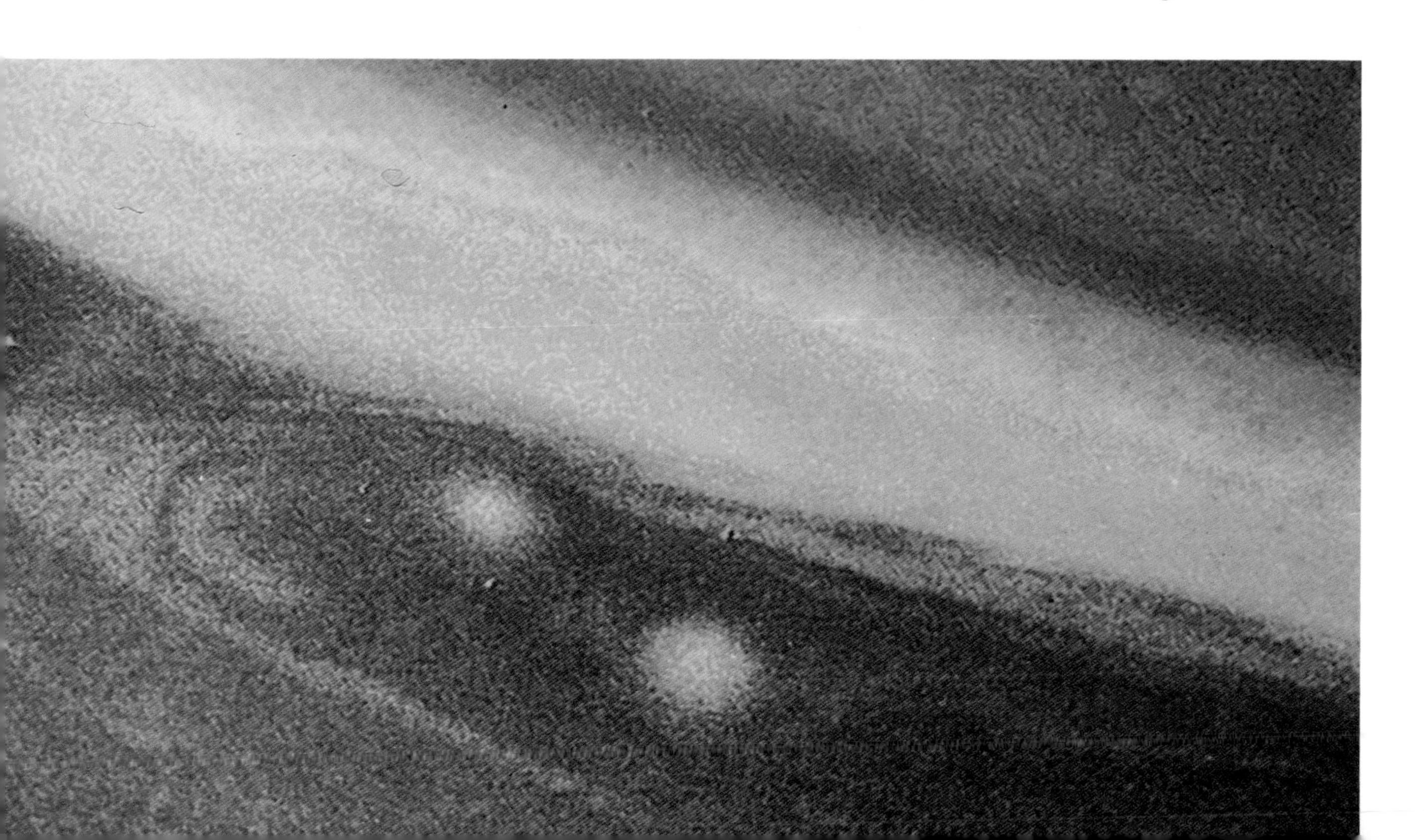

Earth. But we can't burn the methane on Titan, because there's no oxygen in the atmosphere. So our boat doesn't have a fuel tank; it has an oxygen tank instead.

Unfortunately, many scientists now think that Titan may be just a few degrees too warm for the methane to condense as rain and snow. Even so, its surface may not be bare. The orange material of its clouds may be able to coagulate and drip on to the surface far below. We don't know precisely what these orange clouds are made of, but they are almost certainly some kind of organic material, containing nitrogen atoms. Ultraviolet radiation from the Sun can weld together the nitrogen gas in the atmosphere with the methane (and other carbon-containing molecules) to form much larger molecules, and these make up the orange-coloured particles in the clouds.

Chemists first came across this kind of material when they tried to simulate the formation of life on Earth: if you irradiate a mixture of gases like methane and nitrogen, you end up with a waxy orange substance. On the early Earth, this material dissolved in the seas, to become the raw material of the first living cells. Titan, with its water frozen, has probably stopped at an earlier stage. The organic gunge may have rained down on its surface, to build up unchanging deposits of the raw material of life. This planetwide 'experiment', operating over eons of time, should give a much truer idea of how life began than reactions restricted to a glass flask and limited by the period of a scientist's research grant!

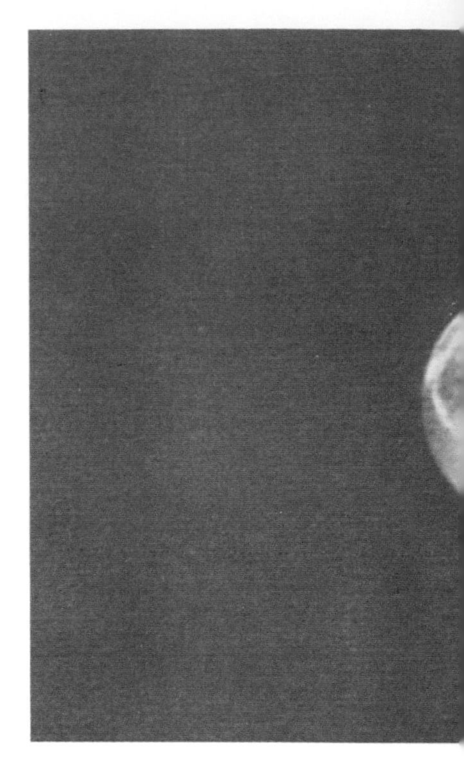

The moon Dione appears in front of Saturn's orange clouds in this picture from Voyager 1. The icy moon has bright streaks across the darker hemisphere on the left; astronomers are still trying to interpret this feature, and many other mysterious aspects of Saturn's moons.

Faced with Saturn's bland globe, and Titan's impenetrable veil, the mission controllers commanded their second Voyager to spend most of its time investigating the planet's rings, and the other moons. Between them, the Voyagers managed to photograph all of Saturn's known moons – and discover so many more that Saturn, with 23 satellites, can lay claim to almost half the moons of the Solar System. The moons are giant cosmic icebergs, made up mainly of ordinary ice, mixed up with only a small amount of rock. Collisions in the distant past have left them covered with craters; Mimas and Tethys each bear a crater one-third as large as the moon itself. Matter has welled up from the interior of some of the moons, painting one half of Iapetus as dark as coal while its other face is snowy white. Dione is daubed with bright streaks of fresh ice, while Enceladus may have sprayed drops of water out into space to spin a faint ring (the E-ring) around Saturn, well beyond the well-known bright rings.

Just outside the biggest of the main rings, the A-ring, Voyager found a strange pair of moons that share the same orbit about Saturn. These gave rise to a false report of a moon called 'Janus' in 1966 (confusingly, one of these two moons has now been named Janus). They are probably two halves of a larger moon that was broken in two by a tremendous impact. The two moons follow almost precisely the same orbit about Saturn, taking almost seventeen hours to circle Saturn once. But one is very slowly catching up on the other. When it reaches it, something very strange happens. Before they can collide, the slower moon picks up speed, and the faster moon slows down correspondingly. The two satellites have now effectively swapped orbits, and they draw apart again. This strange pas-de-deux recurs every four years.

The unexpected behaviour of these satellites provides a clue that gravity doesn't always behave as you expect when you have three bodies (Saturn and two satellites) or more affecting each other. This was the lesson that astronomers learnt the hard way when the Voyagers began to send back pictures of Saturn's rings.

The Voyager scientists were naturally keeping a close eye on the strange F-ring that Pioneer 11 had discovered. As Voyager 1 approached Saturn, it showed the reason for the narrowness of this ring: it lies between the orbits of two tiny and previously-unknown satellites. These 'shepherd moons' corral the tiny ice particles in the ring, forcing them to follow a narrow track. But as so often with Saturn, no sooner was one mystery solved than another appeared.

It came with the first close-up picture of the F-ring: a picture that shocked and mystified. The image on the television screen showed three separate strands to the F-ring; and two were braided around one another. How could icy particles, feeling the same gravitational pulls, follow two different paths that intertwine? That was when Brad Smith declared that his mind was boggled. The known laws of orbits could not explain the F-ring. The next day the media widely misquoted Smith as saying that the F-ring defies the laws of science!

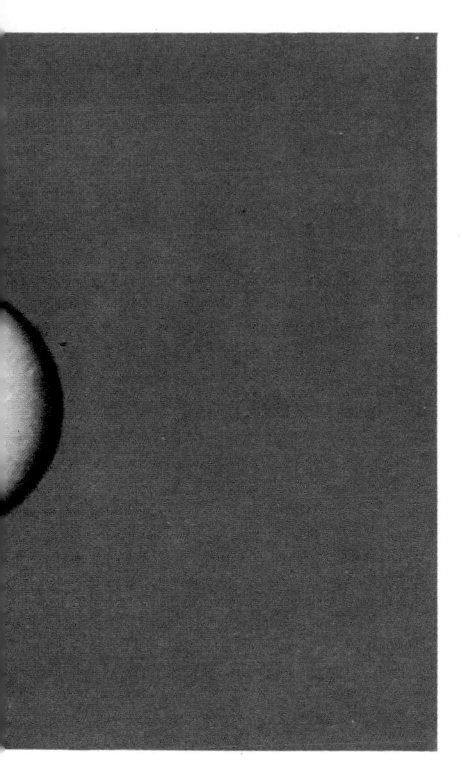

But things weren't that bad. The ring is not twisted all the way around the planet. Voyager 2 took close-ups of as much of the F-ring as it could, and most of it is quite orderly. Meanwhile, the theorists had been calculating what unexpected quirks gravity might produce when we have several bodies to think about: Saturn, the two shepherd moons, and particles of different sizes in the F-ring itself. The two shepherds can affect the small 'sheep' in the ring more than the larger ones, so we might end up with two different paths, one for the smaller and the other for the larger particles. We would see these as two separate strands.

Voyager 1's first views of Saturn's three main rings showed something totally unexpected. Running outwards, across the rings, were elongated dark fingers; and as time went by, they moved round, just like the spokes on a wheel. As Voyager went behind Saturn and looked back towards the dark side of the rings, it saw a negative of its previous view: against the dark rings, the spokes shone brightly. This was a vital clue. Very small particles reflect sunlight in just this way, and so the spokes must consist of tiny ice crystals from the rings, marshalled into spokes by Saturn's powerful magnetic field.

Voyager 1 also showed that Saturn's main rings are more numerous than had previously been thought. As it flew closer to the planet, its improving pictures showed scientists that each ring consisted of many 'ringlets' nestled inside one another. The most detailed pictures revealed 'groove' upon narrow 'groove' separating over 1000 ringlets.

Many scientists wondered if the ringlets were still more finely divided, and Voyager 2 had the chance to find out. It saw Saturn's rings pass in front of the star Dschubba, and the star's light flickered crazily as the rings passed over it. The Voyager scientists found that Saturn has literally tens of thousands of ringlets. Each of them must consist of a stream of icy particles, following each other nose-to-tail around the planet.

Voyager 1's oblique view of Saturn's largest moon Titan shows the orange clouds that totally envelope this world. Above the clouds lie layers of haze (blue).

At first no one could explain the ringlets. Previous theories said that a large gap could be kept clear by the gravity of a nearby satellite. But even with newly-discovered moons, there aren't enough to make thousands of ringlets. The scientists then realized there could be a relatively simple answer. Among all the blocks making up the rings, there could be a thousand or so larger ones that are effectively miniature moons within the rings. These could shepherd the ringlets between them. But Saturn once more delighted in proving the scientists wrong. The flickering light of Dschubba recorded by Voyager 2 never shone through a clear gap between adjacent ringlets: in other words, the dark grooves are not totally empty space between the ringlets. They contain less matter, and appear dark only by contrast.

So Saturn's multitude of ringlets sent the scientists back to their computers. After years of analysing the Voyager pictures, they have concluded that a lot of different things are going on to make the ringlet pattern. For a start, some of the ringlets make a complete circle around Saturn, while others form a tight spiral pattern (exactly like the grooves on a record). Circular ringlets may result just from the natural jostling of the icy particles in the rings, causing 'traffic jams' at regular intervals outwards from the planet. The moons' gravity can stir up the ring particles into spiral patterns. To scientists, it's intriguing that this is the same gravitational effect that creates, on a vastly larger scale, the spiral patterns in galaxies like the Milky Way.

Other spiral patterns in Saturn's rings are more like ripples: the ring bends up and down in a corrugated pattern, about half a mile (one kilometre) high. That is very intriguing. Each time astronomers on Earth have seen the rings of Saturn exactly edge-on, they have tried to measure their thickness. The answer came out as about half a mile. Now it seems that we were just seeing the height of the corrugations, and the rings themselves are much thinner – just as a sheet of corrugated iron seen edge-on appears much thicker than the actual thickness of the iron sheet.

Saturn's rings are probably only 100 feet (30 metres) thick. In relation to their total span of 170 000 miles (272 000km) that is almost indescribably thin. If you wanted to make a scale model out of paper, to show the thinness of Saturn's rings in proportion to its size, you would need to cut out a circle as wide as a city block!

The Voyagers' pictures have refined the idea that the rings consist of separate icy chunks, and now include particles of a wide range of sizes. Saturn's rings must

Voyager 1 took this spectacular picture of the ringed planet as it left Saturn in November 1980. The planet itself is at the upper left, crossed by the narrow shadow of the rings. This close-up shot shows that the rings consist of innumerable thin ringlets. At the left, we can see the globe of Saturn through the rings, showing that they cannot be solid. The large gap in the rings is Casini's Division, and the outermost, isolated, ringlet is the F-ring.

Voyager 1's infamous picture of the F-ring astounded scientists by showing two rings which are apparently braided together, with sharp kinks and bends. It apparently defies the laws of orbits.

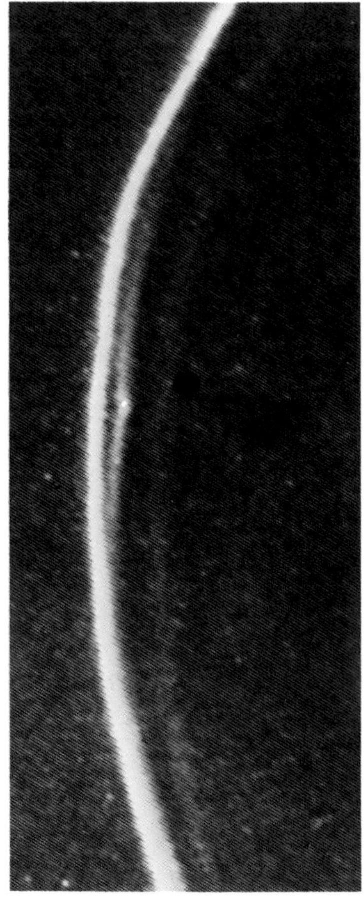

resemble a snowstorm, in which snowflakes are mixed with snowballs and snowdrifts up to the size of a house. In this jumble, the larger chunks of snow are correspondingly rarer, but there are enough of them that an astronaut could move through the rings by pulling on the larger snowballs that are scattered about ten yards (ten metres) apart.

A small snowflake in the rings will eventually collide with a snowball, and stick to its surface. The snowball grows as it collects more flakes and other smaller snowballs, doubling in size in a month. The rings must contain many cosmic snowmen, where a smaller snowball has stuck on top of a larger one! Eventually, the snowballs grow to the size of a small house; and that is a natural limit. Saturn's gravity is pulling slightly more strongly on the side facing the planet than on the opposite side. Small though this difference is, it will eventually tear the fragile moon apart, and disperse its matter as a fresh shower of snow.

This appealing picture fits the facts as we know them today – but Saturn may well have more surprises in store when another spacecraft travels out to probe its secrets. None is yet definitely planned to follow the Voyagers, although American and European scientists are discussing a joint mission that would be called Cassini. This project would have two parts: a study of Saturn's rings, and an investigation of Titan.

Both are important for a study of our origins. A journey to Saturn is a voyage to the birth of our Solar System. The giant moon Titan is an early Earth in deep freeze, where the raw materials of life are preserved for our inspection. And the ring system of Saturn is a microcosm of the birth of the planets. The planets must have been born from the accumulation of smaller balls of rock and ice; and in Saturn's rings the snowballs are still playing and replaying this scene for us. The rings have already taught us that scientists have applied their ideas of gravity too simply; and so many of our calculations on the birth of the planets must be too simplified. Saturn's rings can give us a truer idea of how the worlds of our Solar System came into being.

CHAPTER NINE

URANUS AND NEPTUNE

On the night of Tuesday 13 March 1781, the musician William Herschel despatched the last of his music pupils, and hurried to the garden to follow up the real love of his life: astronomy. Herschel was not just a casual stargazer. He had decided to study every reasonably bright star in the sky, under high magnification. That night, however, held a reward that Herschel could not have foreseen. He became the first person in history to discover a new planet.

William Herschel was born in 1738, as Friederich Wilhelm Herschel, son of a musician in the Hanoverian Foot-Guards. His father taught him the violin and the oboe – and a love of astronomy. The young Wilhelm took up the oboe in the regimental band, but left for England when the Hanoverians were routed by the French at the Battle of Hastenbeck. He became a professional musician, playing the organ in Halifax before ending up as organist and composer at the Octagon Chapel in Bath.

But Herschel also had a burning interest in astronomy. He could not afford to buy the best telescopes of the time, so he taught himself how to make his own. The easiest type to make was a reflector, with curved mirrors made of a metal alloy. On one occasion, the molten metal spilt on to the floor. The flagstones began to crack, and splinters flew in all directions. Herschel's faithful sister Caroline recorded, 'my poor Brother fell exhausted by heat and exertion on a heap of brickbatts'.

In the end, the dedicated amateur astronomer of Bath had made himself better telescopes than the astronomers of the Royal Observatory possessed. Herschel's scientific mind then led him to think what he could do with these superb instruments. His main interest lay in the stars. At that time, no one had measured the distance to a star. Herschel reckoned that if he could find two stars almost in line, the nearer one should appear to move slightly as the Earth travels around the Sun. So, in 1779, the forty-one-year-old Herschel began to look at every reasonably bright star in the sky, to find those that lie in front of a fainter, more distant star. In such a case, the star would appear double when he used a high magnification.

On that evening in 1781, Herschel observed a strange 'star' in the constellation Gemini. He knew that a true star would appear as a point of light, however much he magnified it; if he was lucky, the increasing magnification would show the star as two very close points of light. This new object, however, simply grew in size with the increase in magnification: unlike a star, it was showing a perceptible disc. Herschel noted in his journal that he had found 'a curious either Nebulous Star or perhaps a Comet'.

When Herschel scanned that region of the sky again, the following Saturday night, he recorded, 'I looked for the Comet or Nebulous Star and found that it is a Comet, for it has changed its place.' Herschel did not have the presumption to think he had found a new world, but the Astronomer Royal, Nevil Maskelyne, was prepared to speculate. He wrote to Herschel just two weeks later about this 'comet or new planet'. Within a

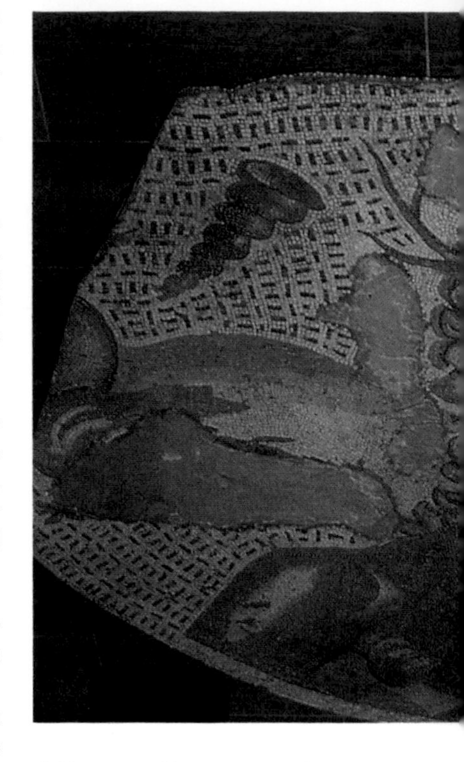

A Romano-German mosaic depicts Uranus, god of the sky. He was the father of Saturn, who deposed him, and grandfather of Jupiter.

Previous page: *Voyager 2 approaches Uranus for its encounter in January 1986. This artist's impression shows the planet surrounded by its nine narrow, dark rings.*

Herschel recorded his discovery of Uranus in the second entry of his log for 13 March 1781, as 'a curious either Nebulous Star or perhaps a Comet'.

Tuesday March 13

Pollux is follow'd by 3 small stars at abt 2' and 3' distance.

as usual. p H

in the quartile near ζ Tauri the lowest of two is a curious either Nebulous Star or perhaps a Comet.

few months, other astronomers had measured the object's motion sufficiently well to show that it was not a flimsy and fleeting comet, but a new world of the Solar System.

The discovery was staggering. Herschel had found an unsuspected planet that was four times larger than the Earth and much more massive (modern measurements put its weight at about fifteen Earths). The planet appears dim because it circles the Sun well beyond the orbit of Saturn; at a stroke Herschel had doubled the size of the Solar System.

Herschel called the new planet Georgium Sidus (George's Star), after the reigning monarch George III. The King made Herschel his personal astronomer, and set him up at a private observatory near Windsor Castle. But astronomers outside England were not so keen on Herschel's name for the new planet. Some called the planet 'Herschel', and the symbol is a circle bearing the letter 'H'. But the other planets have the names of roman deities, and the German astronomer Johann Bode suggested an appropriate mythological name. Working outwards through the Solar System, Mars's father is the next planet, Jupiter; his father was Saturn; and so the planet beyond Saturn should be named for Saturn's father – Uranus.

Although Herschel was undoubtedly the discoverer of the new planet, he certainly wasn't the first to see it. In fact, we can just make out Uranus with the naked eye on a really dark night, as a very dim 'star'. Several astronomers before Herschel had actually marked Uranus as a star on their charts. But they all missed the fact that it was a planet. For one thing, these astronomers were using low magnifications that did not show the planet's disc. Less excusably, they did not notice that it was moving from night to night. In the case of one French astronomer, this could have been because his observations were noted on a paper bag that had contained hair perfume!

Over the decades following Herschel's discovery, astronomers tried to calculate Uranus's path using these older observations of Uranus as a star, as well as their new measurements. But every attempt was doomed to failure. By the 1840s, several astronomers had decided that Uranus was being pulled out of its expected orbit by the gravity of an unknown planet beyond. In principle, it should be possible to calculate where the planet was; but in the days of longhand calculations, that was a difficult and daunting task.

The first to take up the challenge was a young Cambridge mathematician, John Couch Adams. As soon as he graduated, in 1843, Adams began to calculate where the 'eighth planet' must lie in the sky, and for over two years he worked on the problem, in the seclusion of his college. In September 1845, he knew where the planet would be found. He informed the Professor of Astronomy at Cambridge, James Challis, who gave Adams a letter of introduction to the Astronomer Royal at Greenwich. With the country's two major observatories alerted, it seemed that another new planet would be discovered from English soil.

Below: *William Herschel holds a diagram of the newly-discovered planet, in this portrait by J. Russell. The scroll bears the inscription 'the Georgian Planet with its Satellites', and depicts the two moons that Herschel himself discovered (later called Oberon and Titania).*

Below right: *The house in Bath where Herschel was living when he found Uranus is now a museum, decorated as it would have appeared when William and his sister Caroline lived there.*

But due to a mixture of bad luck and misjudgement it didn't work out that way. Adams went to Greenwich, only to find that the Astronomer Royal, George Airy, was in France. When Adams returned the following month, Airy was out during the day; in the evening he was at dinner and the butler told Adams that the great man could not be disturbed. The young mathematician left a copy of his calculations and returned to Cambridge.

George Airy looked at Adams's work, but he was not moved. He wrote to Cambridge, asking if the theory would explain Uranus's changing distance from the Sun, as well as the unexplained motion across the sky. But Adams was, understandably, rather fed up. On top of the fruitless visits to Greenwich, here was the Astronomer Royal asking a question that showed he had not even read Adams's work thoroughly. Adams did not bother to reply; and Airy, supposing the Cambridge mathematician was stumped by his question, did not follow up the prediction. It was stalemate.

Meanwhile, the French astronomer Urbain Leverrier had taken up the same challenge, without knowing of Adams's work. In November 1845, when Airy already had the position of the new planet on his desk, Leverrier published a paper that merely showed that the deviation of Uranus was real, and not caused by the gravity of Jupiter or Saturn. While Airy and Adams remained out of touch, Leverrier completed his calculations. Published on 1 June 1846, his result gave the unknown planet's position in the sky. This time, it was Leverrier's turn to be dogged by bad luck. Possibly because Leverrier was such an unpleasant character, universally disliked by his colleagues, no French astronomer would start a search for his predicted planet.

Airy, the Astronomer Royal, read Leverrier's paper, and compared the French astronomer's position for the planet with the prediction that Adams had left him the previous year. They were almost identical. This could hardly be coincidence, and Airy at last thought it would be worth looking for the new world. None of the Greenwich telescopes was suitable, and Airy knew that the best telescope for the task – by a strange irony – was at Cambridge. Airy had installed this instrument, the great Northumberland refractor, when he had been at Cambridge, before he was appointed Astronomer Royal. Now he wrote to his successor, James Challis, and urged him to begin the search.

Challis was not a man to leap with enthusiasm to this task, or any other. Like Airy, he had had Adams's results on his desk for several months, and had not been tempted even to make a preliminary search with his great telescope. Now he devised a slow and cumbersome strategy. He would measure the positions of 3000 stars in a large region around the planet's predicted position; and, then, several months later, he would measure them again, to see if any had moved. This method was sure and sound, and would certainly turn up the planet, if it existed. (Indeed, a similar painstaking search would eventually net the next planet, Pluto, in 1930.) But it wasn't the quickest way of finding the planet. Challis seemed to have no idea that he was in a race, and that Leverrier's problems with the French astronomers had, for a second time, given him a head start.

In the late summer of 1846, the astronomical world was beginning to thrill with the excitement of the planet-hunt. John Herschel, son of the discoverer of Uranus, was President of the British Association, and he had no doubt that Adams and Leverrier had tracked down a new world. On 10 September he said 'we see it as Colombus saw America from the shores of Spain. Its movements have been felt, trembling along the far-reaching line of our analysis, with a certainty hardly inferior to that of ocular demonstration'.

In Cambridge, Challis continued his painstaking search: but the new planet slipped between his fingers. He actually saw it twice, on 4 and 12 August, and noted its position as a 'star'. If he had compared his observations immediately, he would have found that the 'star' had moved; if he had recalled Herschel's discovery of Uranus by the fact that it showed a disc, he might have tried a higher magnification on the powerful Northumberland telescope, and seen that this 'star' was in fact a planet. Challis's ineptitude is summed up by what happened late the next month. On 29 September, he told his assistant that one star looked as though it might show a disc, but he would wait until the next night before he tried a higher magnification. Apparently the next night started clear, but while Challis chatted about the supposed planet to his

The Cambridge student John Couch Adams was the first person to predict the position of a new planet beyond Uranus; but other astronomers ignored his work.

A French engraving shows Urbain Leverrier engrossed in the calculations that led him to the same conclusion as Adams. Using Leverrier's prediction, German astronomers found the planet Neptune in 1846.

dinner guest, over endless cups of tea, the sky clouded over and Challis lost his chance to observe it.

It was too late. Berlin Observatory had received a letter from Leverrier on 23 September, and the observatory's assistant, Johann Galle, persuaded the director that he should begin a search that very night. They used a telescope only three-quarters the size of the Northumberland, and Galle was at first disappointed that none of the stars showed a clear disc. A young student present, Heinrich d'Arrest, suggested that they should check the positions of the stars against a chart, to see if there was an interloper. By luck, the Royal Academy of Berlin had recently sent the observatory a chart that covered the relevant piece of sky. Peering through the telescope, Galle called out the positions of the stars he saw; d'Arrest checked them against the chart. Before long, Galle called out a star position that led d'Arrest to exclaim 'that star is not on the map'. In one night, they had achieved what Challis had failed to do in two months: they had found the eighth planet.

This time, everyone agreed, the honour of discovering the new planet should not go to the astronomers who first saw it, but to the man who had pinned it down with the power of pure mathematics – Leverrier. The Cambridge mathematician Adams was unknown, and had never published his calculations. England's Astronomer Royal, George Airy, added further insult to his earlier treatment of Adams; although he, of all people, knew that Adams had calculated the planet's position before Leverrier, Airy referred to Leverrier as the 'real predictor of the planet's place'. Fortunately, John Herschel used his considerable influence to let the world know that Adams had been the first to predict the new planet's position, and he arranged for Adams and Leverrier to meet in 1847, in his own house, near Hawkhurst in Kent.

A new planet needs a new name. In the argument that ensued, the hapless Challis saw his failure to discover the new planet followed by total rejection of his suggested name for it, 'Oceanus'; and French astronomers, stung by John Herschel's support for Adams, called the planet 'Leverrier'. The astronomers at the Berlin Observatory, who first saw the planet's bluish-green disc, were keen to call it after the classical god of the sea, Neptune.

The name has turned out to be more appropriate than anyone could have realized at that time. Modern research has shown that Neptune must be made up largely of water. The same is true of Uranus, for the two worlds are twins in size and composition.

Neptune and Uranus are both giants compared to the Earth. It would take four Earths to stretch across the equator of Uranus or the slightly-smaller Neptune; and the heavier of the near-identical twins, Neptune, would outweigh seventeen Earths. But even so, they are considerably smaller than Jupiter or Saturn. Their average density is somewhere between that of Saturn and that of the Earth. So Uranus and Neptune cannot be made of gases, like Jupiter and Saturn; nor of rock, like the Earth and the other inner planets. Only one substance will fit: water. Bill Hubbard of the University of Arizona, has combined all we know about Uranus and Neptune, along with the Voyager's clues about their neighbouring worlds, to come up with the following picture of these twin worlds.

At the centre is a rocky core, weighing about six times as much as the Earth. Surrounding it, and making up most of the planet's bulk, is a deep ocean of hot water. Above this, the planet has a thick atmosphere of hydrogen and methane, which blankets the oceans below. Other researchers have suggested that something more bizarre may be going on. The bottom of the planet's atmosphere may be so hot and compressed that methane gas is breaking up into carbon atoms and hydrogen molecules. The carbon atoms would join together as crystals of diamond; and the oceans of Uranus and Neptune would sparkle with small jewels piling up around the planet's rocky core in a sort of underwater Aladdin's cave larger than the Earth.

Despite their sizes, these two worlds lie so far off that it is difficult to see any details on their discs. As Brad Smith of Arizona's Lunar and Planetary Laboratory has pointed out, Uranus as seen from the Earth appears hardly any bigger than Jupiter's Great Red Spot and Neptune scarcely larger than Jupiter's moon Ganymede. What we have learnt so far, however, shows that the two 'twins' have very different personalities. Uranus is the eccentric introvert; Neptune the hot-blooded extrovert of the family.

Uranus's main oddity is that it lies on its side as it goes around the Sun. Most planets have an axis that is tipped to some extent – the Earth is typical in having a tilt of 23° – but Uranus's axis seems to have been wrenched over by more than a right angle. As Uranus goes around the Sun, it must have the most peculiar seasons. At the moment, the planet's north pole is pointing directly towards the Sun. During the course of Uranus's 'year' of eighty-four Earth-years, the Sun will begin to shine on the planet's equator, on the south pole, on the equator, and then, in 2030, on the north pole again. Holidays on Uranus would be a strange affair. The long periods when the Sun is due south or due north, as seen from the planet, mean that the poles are then the warmest part.

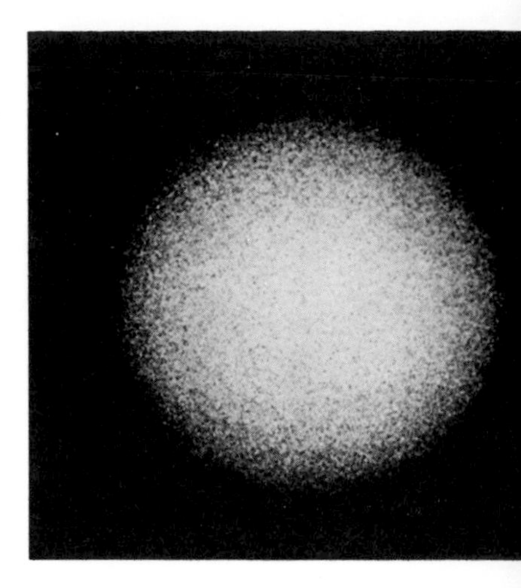

Even the best photographs taken from the Earth show no discernable detail on Uranus. This is partly because the planet is so remote, but it must also mean that the planet has a very uniform covering of cloud.

Uranus also presents a bland, inscrutable face to the Universe. Astronomers who have peered at its small disc, from Herschel to a balloon-borne telescope of the 1970s, have not been able to make out any cloud patterns. The poles are slightly brighter than the equator, but that's all. Because we can't see any details on this world, it's very difficult to work out how fast it is rotating – in other words, the length of its day. The best way is actually to work backwards from the amount that the planet is bulging at the equator as a result of its rotation. The answer is something like sixteen hours.

Unlike Jupiter, Saturn and Neptune, Uranus is not producing extra heat from its centre – or none that we can measure. In fact, its hot-water-bottle of an ocean is probably storing heat from the planet's birth, and the atmosphere is very good at keeping the heat in.

Although Neptune is farther off, and so appears smaller through the telescope, it's a much more interesting world to investigate. White clouds of frozen methane come and go in its upper atmosphere, in bands that run around Neptune between its equator and its poles. Sometimes the clouds spread out to cover the entire atmosphere, making the planet as a whole brighten up. Brad Smith and his colleague Rich Terrile have watched the clouds scudding around Neptune, and have discovered that its 'day' is about eighteen hours long.

Neptune's busy weather patterns are probably due to heat welling up from within. For some reason – possibly its greater distance from the Sun – Neptune has an atmosphere that can't blanket the warm oceans as effectively as Uranus can. Astronomers have used infra-red telescopes to measure the heat trickling through. Although it's less than one-hundredth the heat that Jupiter generates, on this small and cold world any extra source of warmth will make a big difference to the atmosphere.

We'll understand a lot more about the weather on these remote worlds when the space probe Voyager 2 flies past. After leaving Saturn in August 1981, the probe is now *en route* to Uranus: its arrival date is 24 January 1986. The mission controllers have spent the intervening years in rejuvenating the ageing spacecraft. For example, the main radio receiver has become unstable so ground controllers are sometimes uncertain whether Voyager can 'hear' them; and the camera platform tends to stick as it swivels the television camera to follow a target. The platform jammed as Voyager 2 left Saturn, so that frustrated astronomers saw only black sky when they hoped for a sequence showing the 'braided' F-ring in three-dimensions. But a carefully-rehearsed series of picture-taking at Uranus should mean that the camera needs to move only slowly, and the platform should not stick.

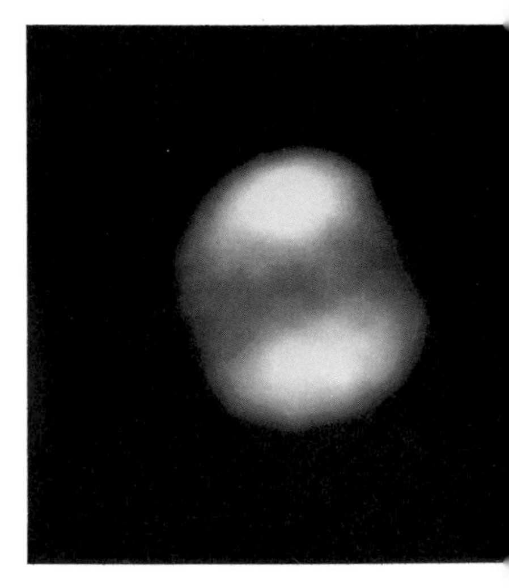

A photograph of Neptune shows distinctly brighter clouds at its poles. These are continually changing, showing that Neptune has much more active weather systems than its 'twin' Uranus.

Voyager's cameras will focus not only on Uranus, but also on its family of five moons. William Herschel discovered the two largest of these icy worlds in 1787, six years after he had discovered Uranus itself. A later English amateur astronomer, William Lassell, saw Uranus's third and fourth moons in 1851.

John Herschel, William's son, named the four moons of Uranus a year later. Other astronomers had rejected his father's idea of naming the planet 'George's Star', in favour of the classical god Uranus, but John managed to break away from the Greek and Roman myths. His names for the planet's moons were taken from English fairy folklore. William Herschel's two moons became Oberon and Titania, the fairy king and queen of Shakespeare's *A Midsummer Night's Dream*. The two fainter moons took their names Ariel and Umbriel from spirits in Alexander Pope's *The Rape of the Lock*. When Gerard P. Kuiper, at the McDonald Observatory in Texas, found a tiny fifth moon almost a century later, he followed the 'airy spirit' Ariel to the pages of Shakespeare's *The Tempest*, and selected the name of the heroine, Miranda.

Because of Uranus's tilt, Voyager 2 will see the planet pole-on as it approaches, with the five satellites' orbits as circles around it – like a cosmic archery target. The probe will miss the bull's eye – Uranus itself – but pass closely inside the orbit of Miranda, obtaining pictures of this small moon as detailed as the best of its views of Jupiter's and Saturn's moons. Because it is meeting Uranus's system flat-on, Voyager has only a short time to take all its pictures. This is where the probe's computer is essential. Uranus lies so far away that radio waves take almost three hours to reach the Earth. In the time that a signal from Voyager 2 takes to reach us, and for our signal to return to the probe, it will have completed its survey of Uranus's system. So the computers have been carefully reprogrammed to do the best they possibly can, with ground controllers taking enormous care not to do anything that might lose control of the distant space probe. Flight engineer Bill McLaughlin says 'when you test a computer on board a spacecraft, you haven't got a second chance – it's like open heart surgery'.

During its brief encounter with Uranus, that planet's gravity will swing Voyager 2 towards its next encounter: Neptune and its moons, in August 1989. Neptune has two satellites. Gerard Kuiper found tiny Nereid the year after he discovered Uranus's small moon Miranda. Nereid follows a large and incredibly elongated orbit that takes it round Neptune in almost exactly one Earth year. Much more exciting is the giant moon Triton, discovered by William Lassell, a Liverpool brewer, a century earlier.

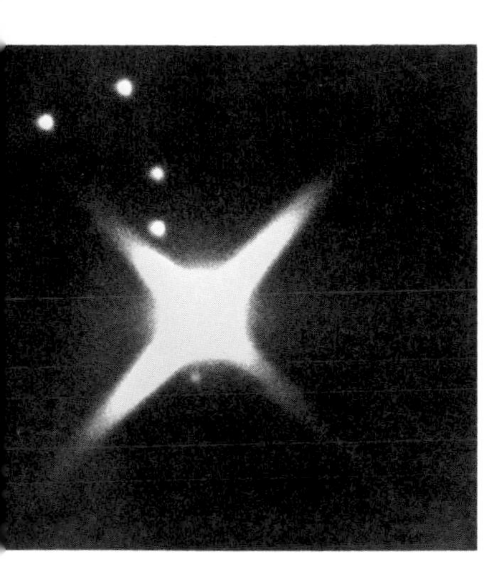

A longer exposure shows Uranus's family of five moons. The faintest moon, Miranda (closest to the planet) was discovered as recently as 1948. (The four bright rays extending from Uranus are not real, but result from reflections in the telescope.)

Some sixty years younger than William Herschel, Lassell followed the famous astronomer's example by building his own reflecting telescopes. In 1844, he completed a telescope with a 24-inch (61-cm) mirror that was then the largest in England. The next year, a friend sent him a copy of John Couch Adams's prediction of a planet beyond Uranus. The story goes that Lassell was laid up in bed with a sprained ankle, and his maid destroyed the letter before he saw it. If Lassell had searched that region of sky with his larger telescope, he would undoubtedly have spotted Neptune as a distant disc – almost a year before the planet's eventual discovery at Berlin.

Once Lassell knew of Neptune's existence, he hurried to observe the new planet. Almost immediately, he saw a faint dot of light next to it, a moon orbiting the planet every six days. It was the moon we now know as Triton, named after Neptune's 'merman' son. It soon became obvious that Triton is a remarkable world. First, its brightness showed that it must be one of the biggest moons in the Solar System.

Voyager 2 is photographing the bluish planet Neptune and its large moon Triton in this artist's impression of the encounter in August 1989. Triton is one of the biggest moons in the Solar System, and observations from the Earth suggest it has an atmosphere and possibly oceans.

Secondly, it goes around Neptune in the opposite direction to the way in which all the other major moons orbit their planets; this is also opposite to the direction in which Neptune itself is rotating. Calculations show that this orbit can't last forever. Just as our Moon is gradually moving away from the Earth, so in Neptune's case, Triton must be drawing closer to the planet. Eventually, its path will spiral right in, and Triton will collide with Neptune.

We cannot measure Triton's size directly, but most astronomers now think it must rank about fourth in the league of satellites – slightly smaller than Jupiter's Ganymede and Callisto, and Saturn's Titan. In the 1940s, Kuiper detected signs of methane on Triton, using the McDonald Observatory's 82-inch (2.1-metre) telescope. Other astronomers have probed more deeply into the light reflected from Triton, and have come up with a variety of portraits.

Dale Cruikshank, of Hawaii's Institute for Astronomy, sees Triton as a world covered by chilly oceans of liquid nitrogen, tinged red by organic substances like those in the clouds of Saturn's moon Titan. Sailing majestically through the red seas are great white icebergs of frozen methane. The McDonald Observatory's Larry Trafton, on the other hand, believes that Triton is a dry world, but that it has a strange atmosphere. During ordinary spells of summer and winter, its methane atmosphere is relatively thin. But every 650 years, Triton's pole is tilted towards the Sun, and a 'heatwave' evaporates the methane ice at the pole. The atmosphere thickens up to something like thirty times its normal density.

When American scientists planned the Voyager 2 space probe, and plotted its Grand Tour of the outer planets, they were looking forward to its views of Uranus, Neptune and their moons, especially Triton. But they did not have a clue that the craft would have something else to investigate when it arrived: rings around Uranus and Neptune.

Jim Elliott had no thought of rings in mind, either, when he boarded a unique flying observatory in March 1977. Elliott wanted to find out more about the atmosphere of Uranus, and he had pioneered a new way of probing distant worlds from the Earth. On the rare occasions when a planet moves in front of a star, Elliott would record the star's light and see how it flickered as the star passes behind the planet's atmosphere. Since 1971, he had explored the atmospheres of Jupiter and Mars using this method. Gordon Taylor, of Britain's Royal Greenwich Observatory, had calculated that Uranus would pass in front of a star in March 1977, and Elliott saw his opportunity to probe this distant planet's atmosphere.

Elliott had a major problem. To be sure of seeing Uranus hide the star, he would have to observe from the region of the south Indian Ocean, where there is little in the way of land, let alone well-equipped observatories. But there was an answer to hand: the Kuiper Airborne Observatory. This C-141 Starlifter transport aircraft is fully kitted out as an observatory, and was named for Gerard P. Kuiper, the astronomer who kept the study of planets alive between the war and the era of space probes. The observatory flies at 41 000 feet (12 500 metres), and astronomers use it mainly to catch radiation that cannot penetrate the Earth's atmosphere to sea-level – infra-red rays that tell of the composition of the giant planets, and of the birth of stars and other planetary systems far off in space. For Elliott, the flying observatory could take him well away from the coast of Australia, and above any clouds that might spoil his view of Uranus's rare passage in front of a star.

As it flew from Perth airport that night, the plane carried a complement of eighteen – a mixture of flightcrew, telescope engineers and astronomers. With all the effort and expense of this trip, Jim Elliott didn't want to miss anything. Forty minutes before he expected the star to disappear behind the planet, he switched on the chart recorder, and its pens began to draw the star's brightness as a pair of red and green lines, wobbling slightly as the star twinkled. And his caution paid off, in a totally unexpected way.

As Elliott and his student Ted Dunham discussed some technical aspects of their observations, Dunham suddenly saw the recorder's pens skid sideways, then back. No one could explain what had happened. It seemed that the star had temporarily dropped to half its normal brightness. Shortly afterwards, it happened again, and again. The team became excited. The effect was obviously real, not just a problem with the equipment. Something out there was passing in front of the star, repeatedly,

American astronomer Jim Elliott shows Heather Couper the chart which first revealed the rings of Uranus. The two traces are a record of a star's brightness as Uranus moved near in March 1977; as the first ring momentarily blocked off the star's light, the pen recorder jumped sideways to make the breaks visible in the middle of the chart.

A 'heat image' provides a direct view of the rings (red) surrounding Uranus. When this infrared picture was taken with the Anglo-Australian telescope, Uranus was pole-on to the Earth, so the rings make a complete circle around the planet. The Earth's atmosphere has blurred the view, so that the radiation from the nine separate rings is merged into one broad band.

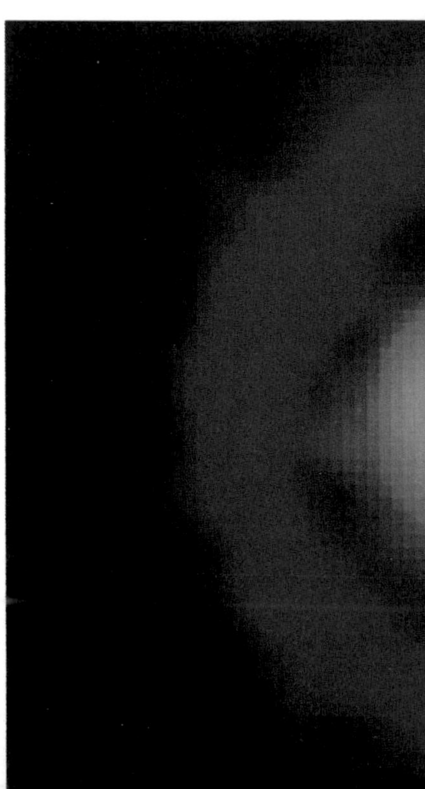

The Kuiper Airborne Observatory is perhaps the most unusual observatory in the world. It consists of a telescope mounted on a Lockheed C-141 transport plane (looking out through the square hatch). In 1977, this telescope first revealed the rings of Uranus, as the observatory was flying high over the southern Indian Ocean.

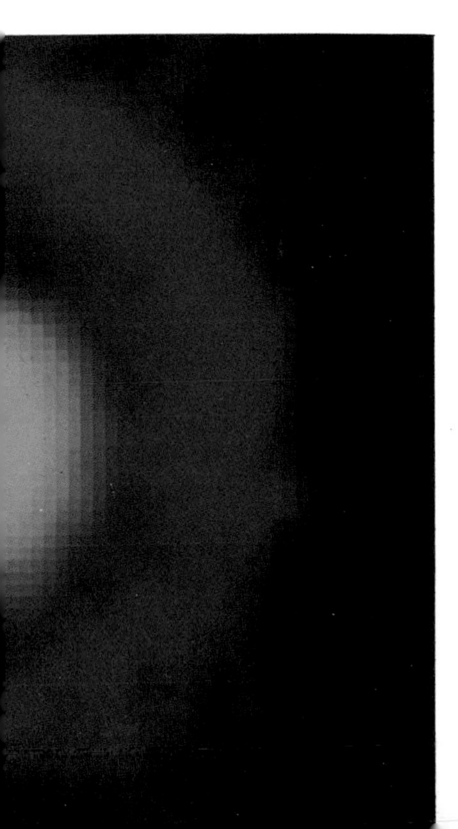

and blocking off its light. In all they noticed five dips on the chart, before the team saw what they had come to get: the star's light fading behind the atmosphere of Uranus.

In the belly of the deafening transport plane, the team had to discuss the matter over headphones, even with colleagues sitting right beside them – and their conversation was recorded for posterity. Elliott joked that the dips in the star's light might be due to 'old satellites that were launched by an ancient race of Uranians', and mentioned, more seriously, the idea of rings around the planet. But the team decided that what they were seeing was a shoal of small moons, all at different distances from the planet, which happened to hide the star.

Once Uranus had passed the star, the astronomers kept the observatory flying until dawn, and picked up similar dips in the star's light when it was positioned on the other side of the planet. Astronomers in Australia and South Africa saw these sudden dips too, even though they were too far north to see Uranus hide the star.

What did it mean? Could Uranus really have hundreds of satellites, all in different orbits? Jim Elliott mulled over the possibilities as he went back to the United States. One far-fetched idea was that Uranus could have thin rings – something unheard of at the time, before the Voyagers had found Jupiter's ring and Pioneer 11 the thin F-ring of Saturn. After dinner one night, he decided to prove that this idea was wrong. If Uranus had rings, then the dips they had recorded on one side of the planet should match those on the other. The paper chart was so long that Elliott had not been able to spread it all out in his Australian hotel room. Now, with the help of his wife, Elaine, he unrolled all eighty feet (24 metres) of the chart on the living room floor. Doubling it over, he lined up a dip on one side of Uranus with one on the other side – and all the others matched too. To Elliott's surprise, his results indeed showed that Uranus has a set of very narrow rings around its equator.

At that time, the only rings that astronomers knew were the broad, bright rings of Saturn. Uranus's rings were the opposite: very narrow, and very dark. Following the discovery, Elliott analysed their results and others in more detail, and found that Uranus has a total of nine rings. All, except the outermost, are only a mile or two wide. At the same time, other astronomers tried to see them directly with telescopes on the Earth but met with failure until 1984, when Brad Smith and Rich Terrile used a new technique to decrease the glare from Uranus itself. The particles making up the ring must be as dark as charcoal. These dark particles do, however, absorb the Sun's heat quite well, so astronomers have been able to pick up infra-red (heat radiation) coming from them.

Neptune has an incomplete ring, forming a series of 'arcs', according to observations made from South America in 1984. This artist's impression shows what Voyager 2 may see when it reaches Neptune.

In early March 1977, Saturn had been the only ringed planet; within two years, Elliott's observations of Uranus and Voyager 1's close-up pictures of Jupiter had added these two worlds to the list of planets with rings. Of the giant planets, Neptune was now the exception. Might it, too, have a ring?

When William Lassell first looked at Neptune and discovered its moon Triton, he thought he saw a ring around the planet – a discovery apparently confirmed by the unfortunate James Challis at Cambridge. But better telescopes showed this ring was an illusion. After Jim Elliott had found the rings of Uranus in 1977, many astronomers looked forward to Neptune's passage in front of a star, so they could discover rings around the outermost giant planet.

The first chance came on 10 May 1981. Jim Elliott had set up instruments in Australia and Hawaii; but to his disappointment, the star showed not a flicker in its brightness before or after the planet itself passed in front. Two weeks later, the planet passed near another star. Again, most astronomers saw nothing unusual – but Harold Reitsema at the University of Arizona's Catalina Observatory did see the star wink off as Neptune passed to the north of it. At the time, it seemed that Reitsema must have been lucky enough to catch an unknown tiny third moon of Neptune passing in front of the star. In June 1983, astronomers all around the world monitored another star as Neptune passed by, including Jim Elliott once again riding the airborne observatory, above Guam in the Pacific. But once more there was no evidence for a ring. Academic papers and popular books both began to refer to Neptune as 'the giant planet without a ring'.

In July 1984, Neptune was due to pass in front of a relatively bright star, and Bill Hubbard, the expert on planets' interiors, was keen to see what information it could provide on Neptune's atmosphere – by that time, he says, 'astronomers had just about given up looking for rings'. He hoped to observe with several different telescopes, but ended up with just two; and no results. His home telescope in Arizona saw only cloudy skies above; while his assistant in South America, Faith Vilas, observed the planet sail clean under the star. But French astronomers were also watching with two telescopes at the European Southern Observatory, sixty miles (100km) north. Both telescopes there saw the star blink out very briefly.

Bill Hubbard, of the Lunar and Planetary Laboratory in Arizona, discovered the arcs of Neptune when he compared American observations with simultaneous results from French astronomers some 60 miles (100km) away. Both were watching a star near Neptune, which 'blinked off' when one of the arcs passed in front of it.

The French team leader, André Brahic, thought they had found another small new moon for Neptune. When he presented the results at an international conference later in 1984, Hubbard became intrigued. He went back to the observations that Faith Vilas had recorded in Chile – and found the same dip in her data. Both the American and the French results showed that the star's light was hidden by something only ten miles (15km) wide; yet it had to be longer than the sixty miles (100km) between the two observatories. This 'object' must be part of a narrow ring going around Neptune. The 'third moon' discovered in 1981 was almost certainly part of the same ring.

Yet the ring cannot stretch all the way round Neptune, because many astronomers had seen a star's light completely undimmed as the planet approached. The ring must be patchy and incomplete. Working together on their results, Hubbard and Brahic agreed that they should call the sections of the ring the 'arcs' of Neptune – partly because the word is the same in French and English. There are probably several arcs following one another at the same distance from the planet. It may be matter that started out like Uranus's narrow but complete rings and is now beginning to clump together. In Hubbard's words, the material 'is trying to decide whether to become a satellite'.

The spacecraft Voyager 2 now has two new phenomena to investigate, undreamt of at the time it was launched: Uranus's narrow rings, and the arcs of Neptune. Both should tell us more about the way that small objects move and clump together as they orbit a larger body, and so reveal more about the way that our Solar System formed. There are no plans for a further probe to either Uranus or Neptune. Given the time taken to plan, to fund, to build, and to fly such a craft, it's likely that Voyager 2 will provide the only close-up views of Uranus and Neptune that we will see in our lifetime.

CHAPTER TEN

PLUTO – AND PLANET TEN

'What, at the moment, is the most distant planet from the Sun?' If anyone asks you this between now and 1999, hold on – it's a trick question. The answer is not Pluto, but Neptune. Although Pluto's average distance from the Sun is more than Neptune's, Pluto has an eccentric orbit that can bring it temporarily – as in the period 1979 to 1999 – closer to the Sun than Neptune.

By the time you read these words, the answer may be different again: we may know of a world beyond both Neptune and Pluto. Astronomers have strong clues that a 'Planet Ten' exists; and they have many new ways of tracking it down. The mid-1980s are seeing a planet-hunt that is unrivalled in the history of astronomy.

Our main clue in tracking down an unknown planet is the effect of its gravity on the other planets. Neptune's pull on Uranus led the mathematicians Adams and Leverrier to predict the position of the unknown planet from observations of Uranus's orbit. This success convinced Leverrier that mathematicians would be able to predict further planets in the same way. He declared 'this success allows us to hope that, after thirty or forty years of observation of the new planet, we should be able to use it in turn for discovering the planet next in order of distance from the Sun'.

In fact, astronomers soon found that Neptune alone could not explain all the discrepancies in the orbit of Uranus; and in the second half of the nineteenth century, they began to search for the culprit. But here, Leverrier's prediction proved to be too optimistic. No one succeeded in finding a ninth planet.

The beginning of the twentieth century saw two brilliant American astronomers working on the problem: William Pickering and Percival Lowell. The two men were at first colleagues, and Pickering helped Lowell to set up his observatory in Flagstaff, Arizona, where Lowell made his study of the Martian 'canals', but they later fell out. Both men began by looking for the unknown planet's effect on Uranus.

In 1908, William Pickering produced his first result for the unknown world he called 'Planet O'. It was about twice as heavy as the Earth, and lay nearly twice as far away as Neptune. Pickering told astronomers that it was then in the constellation Cancer. No one had sufficient faith in these results to take a look. But eleven years later, Pickering included results from Neptune's orbit, and issued a more confident prediction, which placed the world in the neighbouring constellation of Gemini. Astronomers at the Mount Wilson Observatory took this result more seriously. They photographed the region on several nights, and looked to see if any of the 'stars' had moved, hence giving away its identity as a planet. Unfortunately, they did not look hard enough. Although they found nothing, it turned out later that the very faint image of Pluto did lie on four of their plates.

Meanwhile, Pickering's rival, Percival Lowell, was making his own calculations. At first he thought his Planet X was on the opposite side of the sky from Pickering's prediction for Planet O, and Lowell Observatory staff took many fruitless photographs of the constellation Libra. Then, in 1915, he changed his mind. His Planet X was in the region of Gemini and Taurus. Although this was very near to Pickering's position, Lowell had calculated that Planet X was a much more massive world, seven times heavier than the Earth, and only fifty per cent further out than Neptune. As a result, it should have stood out quite clearly on the photographs that the Lowell Observatory took in the spring of 1915. But Lowell was unlucky. Pluto did indeed appear on two of the photographs but it was so much fainter than expected that no one spotted the planet. Disappointed by the failure of his decade-long search, Lowell died in 1916.

The observatory almost died as well. Lowell's wife, Constance, contested Lowell's will, and wanted the observatory to become a museum to her late husband. (So obsessed was she with his memory, that after his death she continued to carry his clothes with her when travelling.) In the ten-year lawsuit the observatory lost a lot of money. But it began a new lease of life in 1927, under Lowell's nephew. He was keen that the observatory should discover 'Uncle Percy's Planet', and he raised the funds for a new photographic telescope.

It would be a tedious task to photograph the sky night after night, and then compare the images of thousands of stars to see if any had moved between one photograph and the next. The observatory's director, Vesto M. Slipher, had to employ a new member of staff – preferably a keen youngster. A young amateur astronomer in Kansas, Clyde

Previous page: *The god of the underworld, Pluto, is a fitting representation for the dim and distant outermost planet. In this Greek statue, he appears with his three-headed dog Cerberus.*

Percival Lowell's adoring wife built a mausoleum to her husband in the grounds of his observatory at Flagstaff, Arizona – in the shape of the planet Saturn.

Opposite: *The rich Boston businessman Percival Lowell built an observatory under the clear skies of Flagstaff to observe the canals of Mars. In later life, he turned to the possibility of finding a planet beyond Neptune – and Pluto was found at the Lowell Observatory, 13 years after his death.*

Tombaugh, had been writing to him with his drawings of Jupiter and Mars and Slipher invited Tombaugh to work at the telescope for a trial period. The young man accepted enthusiastically. Clyde Tombaugh set off for the far southwest, without an inkling of what his new job would entail – or that a bare thirteen months later, he would become the third man in history to recognize a new planet.

Tombaugh had soon mastered the intricacies of the new telescope, and the blink comparator he would use to compare the photographs. The comparator took two photographs of the same region of sky, and showed them in rapid succession. Tombaugh had to set up the photographic plates so that the stars were in the same positions. Then, anything that had moved in the two or three days between the two exposures would appear to jump backwards and forwards, drawing attention to itself.

Slipher, the observatory's director, first asked Tombaugh to take photographs of Gemini, where Planet X should lie. Slipher, and his brother, then took over the blink comparator for several days in a search for a moderately bright 'star' that moved. As Lowell's successor, Slipher felt it only right that he should find the predicted new world. But they saw nothing. Tombaugh could sense their defeat as the Slipher brothers finished the last pair of plates. The irony was that Pluto was indeed on the plates; if the Sliphers had checked every star image, they would have found the planet.

Tombaugh decided on three things. First, since even Vesto Slipher seemed to have lost faith in the calculated position for Planet X, Tombaugh would ignore Lowell's prediction altogether. He would simply search the whole of the Zodiac (the band of sky which the planets appear to follow), and look for any planet beyond the orbit of Neptune. Secondly, he would investigate all the star images on the plates, including the very faint ones, so he could be certain of finding a planet even if it was smaller or more distant than Lowell's Planet X. His third decision was to adopt a systematic strategy. Planets follow looping tracks in the sky, as a result of our moving viewpoint on Earth, and Tombaugh realized that the only way to tell a planet's distance directly from its motion is when it is directly opposite to the Sun in the sky.

So Tombaugh's search fell into place. Every clear night, he was at the thirteen-inch telescope, photographing the next section of sky; by day, he was at the blink comparator, scanning the plates. Many times, he suspected that a faint object had moved, and he checked it on a third photograph, to find it was only a defect on the plate. The routine went on, unchanged, for months. His search strategy brought him around to Gemini. Tombaugh's heart sank. 'I had already written Gemini off,' he admits, because of the Slipher brothers' fruitless search there; and to add to his difficulties, Gemini lies in the Milky Way, and the plates were crowded with stars – almost half a million to be checked on each plate.

But Tombaugh was determined to be thorough. On 18 February 1930, he loaded the blink comparator with plates taken on 23 and 29 January. At four o'clock in the afternoon, he suddenly came across a star image that jumped when he blinked from one plate to the other – and just the right amount to be a planet beyond Neptune. He knew instantly that it must be the ninth planet. 'I was terribly excited – I don't think I could ever top that thrill!'

In the next three-quarters of an hour, Tombaugh checked other photographs of that

Pluto appears as a very faint spot of light (arrowed) on the pair of photographs that led to its discovery. Clyde Tombaugh took these pictures on 23 and 29 January 1930; when he compared them in a blink comparator two weeks later, he noticed that this faint image had moved, and by just the right amount to be a planet beyond Neptune.

region, and there was the faint image, moving in just the correct way. 'Then I was 100 per cent certain'. In the most nonchalant manner he could muster, Tombaugh went to the Observatory Director's office, and, 'I announced "Dr Slipher, I've found your Planet X". Slipher rose up, with a tremendous look of elation – and reservation. I found it hard to keep up with him as he went back to the blink comparator.'

For the next few nights, Tombaugh and the other observatory staff checked and rechecked; then Slipher sent out a telegram announcing the discovery. The date chosen was 13 March 1930. It was 149 years to the day since William Herschel discovered Uranus; and had Percival Lowell still been alive, it would have been his seventy-fifth birthday.

In England, the news broke in *The Times* the following morning. Eleven-year-old Venetia Burney was at breakfast with her mother and grandfather in Oxford when discussion turned to a name for the new world. Venetia had learnt mythology at school, and some astronomy; and she suggested it would be appropriate to name the dim and distant planet Pluto, after the king of the underworld. Her grandfather passed the suggestion to Oxford's Professor of Astronomy, who sent a telegram to Slipher at the Lowell Observatory: 'Naming new planet, please consider Pluto, suggested by small girl Venetia Burney for dark and gloomy planet.'

Clyde Tombaugh is always happy to relate the story of his discovery of Pluto, even after 55 years of retelling the events of 1930. He is now Professor of Astronomy at the University of New Mexico, in Las Cruces.

Slipher and Tombaugh in fact received over a thousand letters proposing virtually every god and goddess in Greek, Roman and Scandinavian mythology. In the end, they chose the name 'Pluto' – but only by default. The most popular choice at the Lowell Observatory was in fact 'Minerva', the goddess of wisdom, but that name had already been given to an asteroid. 'Pluto' had one strong factor in its favour: the first two letters are the initials of Percival Lowell, and a joined 'PL' is used as the shorthand symbol for the planet.

Astronomers could now look back through the old photographs and find Pluto's faint image in previous years – including the lost opportunities of discovering the planet in 1915, 1919, and 1929 – so allowing them to compute the new planet's orbit fairly precisely. They discovered that Pluto has the most oval orbit of any planet, bringing it closer to the Sun than Neptune for part of its orbit, but taking it out to a distance 60 per cent greater than Neptune's when Pluto is at its farthest from the Sun. Although Pluto crosses Neptune's orbit, there is no chance of a collision. Pluto's orbit is more tilted than that of any other planet, so that it passes well 'above' Neptune's path.

But was Pluto the predicted world that perturbs Uranus and Neptune? Certainly, the outer part of its orbit is close to that of Lowell's Planet X; and William Pickering had predicted a Planet O that followed a similar path. There was, however, a major problem. Pluto was so much fainter than either Lowell or Pickering had predicted, that it must be a considerably smaller world. Indeed, when astronomers peered at the new planet using even the highest magnifications, they couldn't make out a perceptible disc: Pluto remained a tiny, starlike point of light. Such a small planet could not have a strong enough gravitational pull to affect the motions of the giants Uranus and Neptune. Yet it was the odd behaviour of these two planets that had led Lowell and Pickering to predict Pluto's position and orbit so accurately.

For decades, the debate rumbled on. Was it just coincidence that Pluto's orbit resembled those of Planet X and Planet O; or was there some way that Pluto could be a massive planet and yet appear small and dim? A small but massive Pluto would have to be denser than lead, which seemed highly unlikely. Alternatively, Pluto could have a highly reflective surface, so what we see is just a small image of the Sun reflected from its centre, just as a ball-bearing reflects a spot of light much smaller than itself. What was needed was a way to measure Pluto's gravity directly – and that meant finding a moon in orbit about the planet. Tombaugh and his colleagues at the Lowell Observatory realized the dilemma soon after they discovered Pluto. With their largest telescope, they looked for a moon close to Pluto and found nothing. There the matter rested until the 1970s.

Jim Christy wasn't looking for a moon of Pluto that day in the summer of 1978. His work at the US Naval Observatory in Washington DC involved measuring, as precisely as possible, the positions of stars and planets on photographic plates. The plates were taken with a specially-designed telescope that had been erected in Flagstaff, Arizona – just a few miles from the Lowell Observatory – and Christy had asked the astronomers there to take a series of photographs of Pluto, so that he could work out its orbit with

The US Naval Observatory's specially-designed telescope at Flagstaff can take very sharp and undistorted pictures of stars and planets, allowing astronomers to measure their positions very accurately. A photograph from this telescope (above) first showed that Pluto has a moon – ironically, it was taken only a few miles from the Lowell Observatory where Tombaugh had discovered Pluto itself.

more precision. The steady air above the telescope at Flagstaff, and the telescope's design, produced images that were extremely sharp, and so could be measured very accurately. Unfortunately, the latest pictures of Pluto were apparently not up to standard: the images of Pluto were not round and sharp, and the plates were put in a box labelled 'defective'. Christy was 'wandering about, with not much to do that week', so at the suggestion of his boss, Bob Harrington, he had dug into the box of defective plates and was now looking at a decidedly peculiar image of Pluto: under high magnification, the planet seemed to be pear-shaped.

Christy thought that something must have slipped during the exposure, so that the image had 'trailed' slightly across the plate. Nevertheless, he started to measure the positions of the background stars, the fixed markers against which he would measure Pluto's motion. In the middle of this, he was interrupted by technicians who had come to repair the machine, and the break gave him a chance to think about something rather peculiar. Despite the shape of Pluto's image, the stars appeared as perfectly circular images. If the exposure had been trailed, then all the star images should be the same shape as Pluto's.

There was another possibility. The pear-shape could be the combination of two circular images, one brighter than the other: the brighter image would be Pluto itself, the fainter object a satellite very close to the planet. 'I didn't really believe it,' he admits. 'I went away and got on with something else, and then later thought: I know it's a moon – I'd better do something about it!' That 'something' was to check other photographs, and sure enough, a high magnification showed that Pluto did appear elongated on other occasions – and the elongation of the pear-shape moved gradually around, exactly as a moon would move round a planet.

So Pluto's moon came to light, almost five decades after the planet itself was found. The following day, Christy checked plates taken several years earlier: yes, some of them showed a definite elongation. To his surprise, his colleagues not only accepted his verdict, but became very excited. More important, in just twenty-four hours, Christy was able to solve the forty-eight-year-old riddle of Pluto's mass and its gravitational pull: Pluto has a very low mass, and its gravity is far too weak to affect the distant giant planets Uranus and Neptune. The coincidence with Planets X and O was just that: sheer, lucky coincidence.

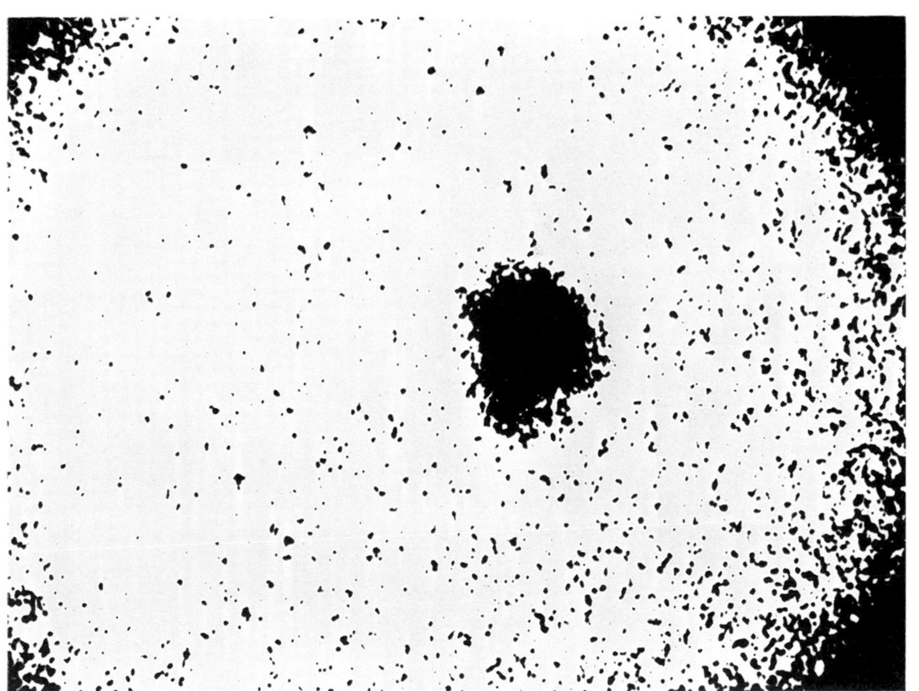

Pluto looks strangely pear-shaped in this highly-enlarged image from a photographic negative. When Jim Christy examined this picture in 1978, he concluded that the odd appearance was due to the image of a moon falling alongside the image of Pluto itself.

The discoverer of Pluto's moon, Jim Christy, named the new world Charon – not after the mythological character, but for his wife Charlene.

As the moon's discoverer, Christy had the privilege of naming it. He had two names in mind. His first choice was 'Oz', from *The Wizard of Oz* – but that was hardly classical enough. The second idea was to name the moon after his wife, Charlene, known as Char to her friends. How about Charon (pronounced Sharon), he mused? Turning to his dictionary, Christy found that Charon was actually a figure in Greek mythology –

and, by amazing coincidence – the ferryman who transported souls to the underworld, the land of Pluto. So astronomers accepted the name 'Charon' for Pluto's moon, envisaging a surly mythological character, while Mrs Charlene Christy is very happy to know that, in reality, it is named for her.

A few astronomers have taken their time to accept Charon as real: perhaps there could be another explanation for the elongated images of Pluto? But new techniques have almost eliminated the amount of blurring caused by the Earth's atmosphere and shown the two bodies separately. The final proof came in 1985. Charon's orbit is very tipped up, even relative to the inclined orbit of Pluto, and only twice in Pluto's 'year' of 248 Earth-years do we get a period when Charon passes directly in front of, and directly behind, Pluto. (It's equivalent to the periods when we see Saturn's rings edge-on.) Harrington and his colleagues predicted that one of these periods would start in the mid-1980s. When Charon passes out of sight behind Pluto, the combined light of the two worlds must temporarily drop to just the brightness of Pluto on its own. Astronomers around the world waited for three years to detect the expected dip in brightness. And on 17 February 1985, Rick Binzel at the McDonald Observatory, West Texas, caught the first dimming. Over the following years, these periodic dimmings should tell us accurately the size of Pluto and Charon; and as Charon moves in front of Pluto, covering and uncovering its darker and lighter regions, the changing brightness should let us construct the first – albeit crude – map of Pluto.

Our best measurements so far tell us that Pluto is by far the smallest planet, no larger than our Moon. Charon's motion around Pluto reveals that, in terms of weight, Pluto lags even further behind the other planets, with only one-thirtieth the mass of the next-lightest planet, Mercury. This means that Pluto has the lowest density of any of the solid planets. It is probably a frozen snowball, a mixture of ordinary ice and frozen methane. Some of the methane has evaporated to form a tenuous atmosphere around Pluto, an atmosphere that should get thicker over the next few years as the planet approaches its nearest point to the Sun, in 1989.

There are no plans for a space probe to visit Pluto in the foreseeable future, but our observations from Earth will be helped when Voyager 2 sends back its pictures of Triton, Neptune's larger moon. Triton is another ice-and-methane world, and is about the same distance from the Sun as Pluto. In imagination, however, we can stand on Pluto's frozen surface, surveying the glittering sheets of methane ice that stretch over the cratered world. The scene is lit only dimly, by a distant Sun that appears no larger than a star, and shines with a brightness that is closer to the light of a full Moon than that of the Sun in our skies.

Charon hangs in Pluto's sky, apparently motionless. In fact, Charon orbits the planet in just over six days, but that is exactly the same as Pluto's day, so that the moon is always above the same spot on the rotating world. (In technical terms, Charon is in geostationary orbit, the same kind of orbit that keeps a communications satellite at apparently the same position in the Earth's skies.) Charon is so close to Pluto that its dimly-lit shape appears several times bigger than the Moon that we are used to. Charon is probably one-third to one-half the size of Pluto itself; and together they constitute what is effectively a double planet.

What lies in the dark recesses beyond the Pluto-Charon double planet? Is there a Planet Ten? Astronomers have speculated about it ever since Clyde Tombaugh discovered Pluto in 1930. In fact, Tombaugh was not convinced that faint little Pluto

Pluto and Charon appear distinctly when they are viewed by a special technique that removes the blurring effect of the Earth's atmosphere. The technique involves computer-processing to give an average image (an autocorrelation), which has given rise to Pluto's unnatural oblong shape. Nonetheless, this picture shows the true separation of the two bodies, and their relative sizes: they are so similar that Pluto and Charon are best regarded as a 'double planet'.

could be rocking the orbits of giant Uranus and Neptune. Soon after he found Pluto, Tombaugh was back at his routine: photography at night, blinking the plates by day. He carried on for another thirteen years, until the Second World War interrupted his search. First he completed his survey of the Zodiac; then in successive years he looked at strips to either side, until he had personally inspected seventy per cent of the sky.

A Victorian engraving depicts the ferryman Charon who ferries the souls of the dead across the River Styx to the land of the dead, the realm of Pluto.

In his extended search, Tombaugh saw more stars than anyone before or since: he individually scrutinized forty-five million stars. His thorough study turned up many new objects in the sky: 775 asteroids; 1700 stars that vary in brightness; six clusters of stars; and several examples of distant galaxies clustered together. But there was no new planet.

After the failure of this heroic enterprise, astronomers put the question of Planet Ten to one side for some forty years. But the 1980s have seen a tremendous revival of interest. Most astronomers are convinced there must be another world out there; and many of them are looking for it, spurred partly by the immortal fame the discovery will bring. There are good reasons for this sudden enthusiasm. Jim Christy's discovery of Charon in 1978 proved, once and for all, that the strange motions of Uranus and Neptune are not caused by the gravity of Pluto – so, there must be a world beyond that which is responsible. In the 1980s, we have better signposts than Lowell and Pickering could use. And finally, new kinds of telescope, and computer processing, can find the new planet much more easily than Tombaugh's blink comparator.

Uranus and Neptune still provide the starting point. Since the time when Lowell and Pickering based their calculations on Uranus, the planet has completed another, accurately-measured orbit about the Sun, making a total of three-and-a-half orbits since it was first unknowingly catalogued as a star. And Neptune has now gone round the Sun once since the French astronomer Joseph Lalande measured its position as a 'star' in 1795. In a fascinating piece of historical research, Stillman Drake and Charles Kowal have found that Neptune was seen earlier still – by Galileo! In January 1612, Galileo was charting the positions of Jupiter's moons, when this planet happened to move across in front of Neptune. Galileo's notebooks clearly show Neptune as a star; and, because we can calculate Jupiter's position accurately, we can tell where Neptune was. There's still some dispute about how accurately Galileo marked Neptune's position, but it does give us an idea of where the planet was more than two orbits ago.

The perturbations of Uranus and Neptune are the most solid evidence we have for Planet Ten. But the gravitational effect of this very remote world on the two planets is diluted by its distance, and we could get a much better idea of Planet Ten's position if we could track several objects that go farther out in the Solar System, and cover quite a large region of its gravitational field. For a century, astronomers have been trying to use comets in this way. Comets are small icy bodies that come in from the outer reaches of the Solar System, and they respond very readily to the powerful gravitational tug of a giant planet. Early this century, William Pickering analysed the motions of comets, and predicted several giant worlds very much further out than his Planet O. Distant though they are, Tombaugh would have picked up these large planets in his thorough survey, and we can be sure they don't exist. More recently, a special search was set up for a supposed planet that was affecting Halley's Comet, but again the planet was never seen. The problem with comets is that they are not totally predictable. A comet, more often than not, is ejecting jets of gas that propel it away from the path we would expect.

But now we have four artificial 'test bodies' going to explore the gravitational fields of the outer Solar System: Pioneers 10 and 11, and Voyagers 1 and 2. The redoubtable Pioneer 10 is now the most distant known object in the solar system. Launched thirteen years ago, it is still going strong, and has now passed well beyond all the planets. If nothing goes wrong, NASA should be able to 'hear' the faint signals from Pioneer 10 until after the year 2000, when it will be at twice Pluto's average distance from the Sun.

Although the NASA mission planners didn't intentionally send the four probes to track down Planet Ten, they are the ideal scouts. As they head outwards, their paths will be gradually bent by the gravity of any planet beyond Neptune and Pluto. They are sufficiently spread out to experience different gravitational pulls from the planet, helping to pinpoint its position; in particular, Pioneer 10 is heading in the opposite direction from the other three. It's also much easier to work out the position and speed

The redoubtable spaceprobe Pioneer 10 is now beyond all the planets, heading for the emptiness of interstellar space. If there is a planet beyond Neptune and Pluto, Pioneer 10 should soon be feeling its gravitational tug.

of a spacecraft than a planet, because it is emitting a radio broadcast of a precisely-known wavelength, and radio telescopes can pin down positions much more accurately than any other kind. So far, the space probes have not moved noticeably away from their calculated paths: but at any time, one may begin to deviate, and, once it does, that effect can but grow until it points a finger towards Planet Ten.

Tombaugh's disappointing thirteen-year search for Planet Ten sets important limits on the kind of planet we can expect to find. Unless its orbit is very tilted, and the planet lies a long way from the Zodiac in the region Tombaugh didn't search, the planet must be more than ten thousand times fainter than the dimmest stars visible to the naked eye. This means that there cannot be a large planet just beyond the orbit of Pluto. Tombaugh's lack of success means that there is not another planet like Neptune, for example, at less than seven times Neptune's distance from the Sun.

If Planet Ten exists, however, it's almost certain that our telescopes have already recorded it. Its image may be sitting on long-exposure photographic plates, or it may be recorded as an object in a computer catalogue of sources of radiation in the sky. Some astronomers are not looking for obvious signposts to Planet Ten, but are simply

sifting through mounds of astronomical data, hoping that the new planet will give itself away, perhaps by its gradually changing position.

Since the 1950s, astronomers have used a new kind of astronomical camera, the Schmidt telescope, to take wide-angle pictures of the sky. Several Schmidt telescopes around the world have now photographed the entire sky, recording the images of thousands of millions of stars. The faintest are a hundred times dimmer than the faintest on Tombaugh's photographs. No one could 'blink' so many images; and these days, no one needs to. A laser-scan can convert the photographic image to electrical signals, which can then be processed in a computer. A software programme can then 'look' for a star image that moves. Modern electronics has meant that no one is ever likely to beat Tombaugh's epic 'blinking' marathon.

Planet Ten could be too distant or faint to appear even on these photographs, but it was undoubtedly spotted – if it exists – by a unique refrigerated satellite. The Infra-red Astronomical Satellite (IRAS) was launched in 1983 to search for sources of heat (infra-red) in the Universe. From its viewpoint, it was untroubled by the Earth's atmosphere which absorbs infra-red rays from space; and a surrounding tank of incredibly cold liquid helium prevented the telescope from being swamped by its own

Above right: *The 210-foot (64m) antenna of NASA's Deep Space Network at Goldstone in California is picking up the faintest whispers from Pioneer 10 as it heads farther into space. This giant dish (note the figure by the doorway for scale) will be able to track the spacecraft into the next century, until it is twice as far away as Pluto.*

heat radiation. IRAS had to avoid looking at the Sun or the Earth, whose powerful heat radiation would have destroyed the detectors. In fact, all the objects in the solar system, warmed by the Sun's heat, were very bright as seen by IRAS's infra-red eye. A planet beyond Neptune and Pluto would not be warm by everyday standards – but the supercooled IRAS would have picked up its heat.

IRAS discovered, or suspected, over half a million sources of heat in the sky. Its catalogue is stored on a computer tape, and also published as a pile of computer printout some nineteen inches (48cm) thick. If Planet Ten exists it is an entry in this catalogue. Many astronomers have tried to pick it out; so far no one has succeeded.

Not everyone is even convinced that the perturbations of Uranus and Neptune are caused by a planet. Mark Bailey of Sussex University thinks that comets are responsible: beyond the planets lie a large swarm of comets, and if they are densely packed at one point, this clump of comets could provide the gravitational effect that we measure. More popular is the idea that a 'dark star', more massive than any planet, is lurking well beyond the planets that we know; it would be the 'Nemesis' that some astronomers hold responsible for the death of the dinosaurs.

The Infra-red Astronomical Satellite (IRAS) surveyed the whole sky for sources of heat, and it must have 'seen' Planet Ten – if it exists. This artist's impression shows the American-Dutch-British satellite above its control centre in England. IRAS was built around its telescope, which looked through the circular hole at the top; much of the rest of the satellite consisted of a cooling system to refrigerate the heat-sensitive telescope.

But one astronomer who's convinced that Planet Ten exists is Bob Harrington, of the US Naval Observatory. On his desk, he has a card which casual visitors may not study in detail. The card bears the orbit of Planet Ten, as calculated by Harrington and his former colleague Tom van Flandern.

Harrington and van Flandern believe that Pluto's odd orbit, the backwards orbit of Neptune's big moon, Triton, and Planet Ten are all part of the same puzzle. Neptune once had a system of moons that went around the planet in the same direction as the planet's spin: among the moons were Triton, and Pluto. At some point in the past, Planet Ten swept closely past Neptune, reversing Triton's orbit and throwing off Pluto as a separate planet. In the violent encounter, the original Pluto split into two parts. Pluto and Charon. To make this encounter work out, Harrington and van Flandern found that the errant planet had to be as heavy as four Earths, and to travel round the Sun in an orbit that is highly tilted, very elongated and gives the planet a 'year' of 800 Earth-years. When they had done this calculation, Harrington was amazed: 'You get exactly the same planet as you need to explain the perturbations of Uranus and Neptune.' This convinced them that they were on the right track.

Clyde Tombaugh would have seen this planet if it had appeared on his plates, so Harrington surmises that its tilted orbit took the planet so far from the Zodiac that it fell in the region that Tombaugh abandoned because of the war. But it would now be heading back towards the Zodiac, and Harrington's colleagues are beginning to take photographs to try and catch the planet. Identification may come another way, Harrington is convinced 'IRAS has got it'.

If Harrington is correct, he'll have the first choice of a name for the world. It's something he hadn't given any thought to until a journalist asked him what name he would choose. As befits a colleague of Jim Christy, who named Pluto's moon after his wife, Bob Harrington is not a man to take a heavy classical approach. He thought for a second, and drew a name from the air. And so the grand-sounding Humphrey Project of the august US Naval Observatory is not named for an eminent astronomer, politician or admiral. It is the unofficial title of Harrington's search for the missing world of our Solar System: planet Humphrey.

Bob Harrington confidently waves the computer card which tells where Planet Ten lies – according to his calculations.

CHAPTER ELEVEN

ASTEROIDS AND COMETS

The 'Great September Comet' of 1882 was clearly visible in the dark skies of Egypt. The comet's elongated tail was temporarily the largest object in the Solar System, but made of extremely insubstantial gas and dust. This great show was, ironically, created by one of the Solar System's smallest bodies, an icy block only a few miles across which has come so close to the Sun's heat that the ice has boiled off as gas.

Just as everyone is brought up knowing the seven days of the week, so schoolchildren know that there are nine planets going around the Sun. When we give a lecture in a planetarium, we can always rely on the children's voices to chant from the darkness of the artificial night: 'Mercury, Venus, Earth . . . Pluto'.

But the nine planets are not the sum total of the Solar System. Astronomers have charted, and named, thousands of smaller worlds – and know that there are millions more, too small to follow individually. Some of these worlds are in orbit about the planets: the larger ones are the planets' moons, while countless smaller ones make up the rings of Saturn, Uranus and Jupiter, and the arcs of Neptune. Other small worlds orbit the Sun as 'minor planets' in their own right, mainly in a wide band between Mars and Jupiter. And astronomers believe there is a swarm of small icy worlds that orbit the Sun, way out beyond the known planets.

Until recently, astronomers didn't pay much attention to the smaller worlds. For a start, they only amount to somthing like 0.02 per cent of the matter orbiting the Sun – the overwhelming majority being locked up in the nine planets. Then, there was the practical point that telescopes on Earth can't show these tiny, distant worlds as anything but points of light, so we could not hope to learn much about them. Even the promise of space probes created little interest. Most astronomers believed that anything smaller than our Moon would be even less interesting than the Earth's dead companion.

How wrong we were! Astronomers' eyes were opened when the Viking spacecraft sent back their pictures of Mars's two moons, Phobos and Deimos; and especially when the Voyagers photographed the strange worlds that orbit Jupiter and Saturn. Smaller bodies are not just miniature versions of our cratered Moon: they bear grooves, huge splashes of dark or bright deposits, refrozen seas; and even thick atmospheres or volcanoes. The active moons are obviously exciting in their own right; but astronomers are now taking an interest in the less-active moons, too, because their unchanging surfaces are relics of the formation of the Solar System.

Astronomers have come to realize that small icy worlds winging inwards from the depths of the Solar System are responsible for the most spectacular objects in our

Previous page: *Like a celestial peacock, Comet West displays its colourful tails in the morning skies of March 1976. The narrow blue tail is made of glowing gases; the broader yellow tail of small dust particles.*

skies: the comets. The awesome spectacle of a huge glowing head, and a tail hundreds of millions of miles long, is merely a vast lightshow generated by a dirty snowball. The snowball itself, at the comet's centre, is only a few miles across – so small that our telescopes cannot reveal it hiding in the centre of its glorious display of gases.

Astronomers now make less of a distinction between moons, asteroids, and comets. Mars's two moons, and some of the satellites of Jupiter, are probably very similar to the minor planets – they may well be captured asteroids, in fact. If we took one of the icy moons of the outer planets, and brought it close to the Sun, its matter would evaporate to surround it with a comet's head and tail. What seems most important is a minor world's distance from the Sun. The asteroids are so close that any ices have long since evaporated away; they are rocky bodies. The satellites of the outer Solar System, and the comets' dirty snowball nuclei, lie so far away from the Sun's heat that they are made mainly of ices.

It is also less easy to draw the dividing line between these smaller worlds and the nine 'proper' planets. Jupiter's largest moon, Ganymede, is larger than Mercury, and considerably bigger and more massive than Pluto. Only because Ganymede orbits Jupiter rather than the Sun is it not classed as a planet. Now that we know Pluto is a very small and lightweight planet should we regard it instead as a large asteroid? After all, the biggest asteroid, Ceres, is closer in size to Pluto than Pluto is to the Earth.

The story of the asteroids starts before astronomers had any idea of planets beyond Saturn. Kepler's laws of planetary motion gave astronomers a scale model of the Solar System, and two things were immediately obvious. The spacing between the planets becomes greater and greater as we look at planets further from the Sun; and there is an unexpectedly large gap between Mars and Jupiter. In 1596, Kepler remarked, 'Inter Jovem et Martem interposui planetam' – 'between Jupiter and Mars I would put a planet'. 170 years later Johann Daniel Titius, a German mathematician, found a formula to describe the planets' distances from the Sun. Johann Bode, director of the Berlin Observatory, took up Titius's formula enthusiastically, because it predicted the precise distance of the 'missing planet'.

The Titius-Bode Law (sometimes known, rather unfairly, as Bode's Law) is surprisingly simple. It is a cosmic version of the 'spot the missing number' game. If we call the Earth's distance from the Sun 10 units, then the six planets known at the time – Mercury to Saturn – had distances of approximately 4, 7, 10, 15, 52 and 95 units. The big gap comes between 15 (Mars) and 52 (Jupiter). Titius found the formula to link these numbers. Take the series 0, 3, 6, 12, 24, 48, 96, where each number (after three) is twice the preceding number. Add four to each. The series becomes 4, 7, 10, 16, 28, 52, 100. This is surprisingly close to the actual distances of the planets from the Sun (Mars and Saturn are both slightly out, but only by a small percentage). And notice the '28' in the series. Even without understanding why the planets obeyed the series, Bode took this as proof there must be a planet at this spot, between Mars and Jupiter.

At first, other astronomers were not so convinced. But, nine years later, William Herschel stumbled across the distant world Uranus (a name actually proposed by Bode). In terms of the units we've been using, its distance is 192. The next number in the Titius-Bode series is 196. It was too close to be coincidence; and Bode and his colleagues organized a 'Celestial Police' of both professional and amateur astronomers to track down the missing world.

In fact, they were forestalled. An Italian astronomer working in Sicily, Giuseppe Piazzi, came across a small world in just the right orbit, on the first night of the nineteenth century, 1 January 1801. Piazzi was trying to measure the extremely slow motions of stars across the sky, and had installed a telescope that allowed him to measure the stars' positions very accurately. In January 1801, he found that a 'star' he had recorded on the first day of the month was moving rapidly, from night to night. Its motion showed that it must be a world orbiting the Sun, between the orbits of Mars and Jupiter – and at the predicted distance of 28 units. Piazzi named the world Ceres, after the patron goddess of Sicily.

By the time news of Piazzi's discovery had made its way to the Celestial Police in Germany, Ceres had moved into the Sun's glare and could not be seen. Fortunately, one of the greatest mathematical geniuses in history, Carl Friedrich Gauss – then only twenty-four years old – devised a new method of calculating orbits from just a few observations, and used Piazzi's results to predict where the little world would reappear on the other side of the Sun. Almost a year after Piazzi's discovery, Heinrich Olbers, one of the most active of the Police, rediscovered Ceres. Several months later, as he followed Ceres round its orbit, Olbers discovered another small moving point of light. It was a second small world, travelling in an orbit between Mars and Jupiter. The Police were intrigued. They had set out to catch a new planet; now there appeared to be at least two accessories instead. Pursuing their investigations, they found more. Olbers followed his discovery of the second object, Pallas, with Vesta in 1807; and Karl Harding found Juno in 1804.

A small meteorite, cut open for analysis, displays an interior made of several types of minerals and an outer 'fusion crust' where its surface has melted during its fiery fall through the atmosphere. Although most meteorites are fragments from asteroids, the composition of this specimen indicates that it was once part of the Moon's surface. The rock was probably ejected into space when a large meteorite hit the Moon and gouged out a crater.

What should astronomers call these new objects? William Herschel looked at them with the highest magnification he could on his giant telescopes, and could hardly see a perceptible disc, even for the largest, Ceres. Because of their appearance, he proposed the term 'asteroid' (starlike). Other astronomers preferred 'minor planet'; and both terms are still in use today.

For almost four decades, the number of asteroids remained the same. Astronomers needed new and better star maps in order to locate the tiny moving specks of light, and these only became available in the 1840s (we have already seen that one of these new star charts allowed Galle and d'Arrest at the Berlin Observatory to discover Neptune in 1846). In 1845, Hencke, a postmaster and amateur astronomer, found the fifth asteroid. By the end of 1891, astronomers had found 323 asteroids; the next year, Max Wolf at the Heidelberg Observatory introduced a new technique of photography which speeded up the search tremendously. If you take a long exposure of the sky, tracking the stars so they appear as dots, a moving asteroid appears as a short streak. The total passed 1000 in 1923, and astronomers began to regard asteroids merely as 'vermin of the sky' appearing as spurious stars in short-exposure photographs, and spoiling long exposures with their streak across the plate. Once computers arrived, it was easier to keep track of all the asteroids, and the numbers shot up again: at the begining of 1985, there were 3200 asteroids with known orbits.

The person who keeps track of this myriad of worlds is Brian Marsden, of the Minor Planet Center in Cambridge, Massachusetts. His VAX computer holds records of 60,000 individual measurements of the positions of asteroids. It also remembers the orbits of the three thousand asteroids whose paths are accurately known, and at the touch of a key, the computer will produce the positions of all these worlds for any date and time you care to name. When someone reports a new asteroid, Marsden calls on the VAX to list all the known asteroids in that region of sky. If the asteroid is indeed new, then Marsden needs sufficient information to calculate an orbit, accurately enough that astronomers can pick it up again when the asteroid has passed behind the Sun. The VAX takes only a fraction of a second to make the calculations that Gauss slaved over for months when he first tried this technique on Ceres in 1801.

Once an asteroid's orbit is properly worked out on the VAX, it is given a number, and a name. With so many asteroids already known, it's quite a problem to keep dreaming up new names! At first, astronomers followed Piazzi's example in naming Ceres, and called the minor planets after classical goddesses. Fairly soon, however, they had worked their way through the ranks of the goddesses, and of classical heroines; and began to use more familiar female names. Asteroids also became a way to immortalize terrestrial places and events in the sky: Bohemia, Vienna, Arizona and Chicago have found their way to the asteroid belt, as has the liberation of France after the Franco-Prussian war in 1871, as the minor planet Liberatrix. There are now also minor planets Herschel (the two-thousandth to be discovered), Einstein, Marsden and Moore – named for Britain's leading astronomy popularizer.

When Olbers found the second asteroid, he suspected that the two bodies were part of a planet that had exploded. Even today, many Soviet scientists believe this theory. But it does present problems – not least, how could a stable body blow itself apart? – and most Western astronomers accept the opposite viewpoint. The asteroids are fragments left over from the birth of the Solar System. The strong gravity of nearby Jupiter wrecked their chances of becoming a world, stirring up the asteroids' paths as they tried to come together. Jupiter probably stole much of the matter of the still-born planet, too, because what's left in the asteroid belt would make up a planet less than one-tenth as massive as our Moon.

Nature has kindly arranged that we have some samples of asteroid material delivered

free to our doorstep – the meteorites. Meteorites are solid bodies that plunge through the atmosphere, and survive their fiery passage to hit the ground as solid objects. The biggest meteorites strike with such force that they gouge out craters, like Arizona's Meteor Crater. Such a meteorite is of little use to science, however, because the tremendous impact boils away most of the meteorites's material as vapour. Much more useful are the smaller meteorites which land more or less intact. There are three main types. Some are a very dark rock, their black colour coming from carbon; others are more ordinary rocks; and the third type is composed of metal – an alloy of iron and nickel. Although a few meteorites probably come from the Moon or from Mars, the vast majority originate in the asteroid belt: they are splinters from the minor planets.

Different asteroids reflect light and infra-red in a way that is highly reminiscent of the different kinds of meteor. Among the largest asteroids, we can therefore expect a diversity of worlds as different as the four main moons of Jupiter. Ceres is the 'giant' of the asteroids, just over 620 miles (1000km) across, and containing one-third of all the matter in the asteroid belt. Its surface is covered with dark, carbon-rich clays and it may be 'iced' with a thin coating of frost. Pallas and Vesta come next, each about 350 miles (560km) in size. Pallas seems to be a smaller version of Ceres. But Vesta is a lighter-coloured world: although only half the size of Ceres, it reflects so much sunlight that it is the brightest of the minor planets – in fact, you can pick out Vesta with the naked eye, if you know just where to look. Vesta's rocks seem to have melted and covered its surface with flows of lava that are lighter than the original carbon-rich rocks.

Although no one has yet built an asteroid probe, we may get our first view of a stony asteroid in just a few years' time. When NASA mission planners plotted the course of their Galileo probe to Jupiter, they found it passes very near to an asteroid called Amphitrite, a minor planet some 130 miles (200km) across. After Galileo's launch in 1986, the mission planners hope to put Galileo on a path that will send it close by Amphitrite. Its cameras will give us a view as good as the Voyagers' pictures of Saturn's moons.

Meteor Crater, Arizona, was blasted out by the impact of an iron meteorite that hit the Earth some 50000 years ago. The crater is an immensely impressive natural phenomenon on the human scale (as represented here by Nigel Henbest), but it is puny in comparison with the immense craters of the Moon and the other planets.

We expect that a bright stony asteroid like Vesta or Amphitrite has melted inside, and its content of iron and nickel should have dripped to the centre to form a core like the Earth's. In the asteroid belt, the small worlds are constantly colliding, and chipping off each other's outer rocks. As a result, some minor planets have lost all their rocks, and are just huge chunks of metal in space. The largest of these iron worlds is Psyche, a lode of pure iron-nickel some 150 miles (250km) across in space. In the twenty-first century, the world's need for metal may be met, not by plundering our already-ravaged planet, but from a captured iron asteroid in orbit above the Earth.

Although most asteroids lie within the main belt, between the orbits of Mars and Jupiter, we do find some beyond. Two groups are in fact in the same orbit as Jupiter, one leading the planet and the other trailing behind. These satellites are named for heroes of the Trojan wars – the Greeks ahead of Jupiter, and the Trojans behind. The largest of the Trojans, Hektor, is a strange world, shaped like a peanut. Bill Hartmann and Dale Cruikshank have surmised that Hektor was once two asteroids, in close orbit around one another, and that they have spiralled together until they touched and stuck, like two eggs glued together at their pointed ends.

What lies further out in the Solar System? Charles Kowal has been using the Palomar Observatory's big Schmidt telescope (essentially a sensitive wide-angle camera) to find out. Like Clyde Tombaugh fifty years earlier, Kowal looks for objects that move from night to night: the difference is that his telescope can photograph objects a hundred times fainter than Tombaugh's instrument could spot. The big prize for Kowal would be Planet Ten; but he also expects the search to turn up other unexpected objects in the outer Solar System. He was not disappointed. On 18 October 1977, he found a new world, orbiting the Sun between the orbits of Saturn and Uranus.

Kowal named his discovery Chiron, after the wise centaur who was son of Saturn and grandson of Uranus. At first, some astronomers called it a 'mini-planet'. But Chiron does not rank as a planet: with a diameter of only 200 miles (300km), it is only one-third the size of Ceres. Nonetheless, it does share some characteristics with Pluto: both are comparatively small worlds in a region of giant planets, and both pursue strangely elongated orbits. Bob Harrington and Tom van Flandern, who propose that Pluto was a satellite of Neptune flung out by the close passage of Planet Ten, suggest that Chiron was another moon of Neptune, this time sent in towards the Sun by the encounter. Other, more orthodox, astronomers believe that Chiron and Pluto are two members of a second asteroid belt that lies in the outer Solar System: so far we have found only the largest of these asteroids, Pluto, and the closest to the Sun, namely Chiron.

Unlike Pluto, however, Chiron cannot stay for ever in the solar system. Within a few million years – a short time, astronomically speaking – the gravity of Jupiter or Saturn will fling Chiron out of our planetary system, and into interstellar space. The shape of Chiron's orbit, and its impermanence, have suggested to many astronomers that Chiron is not an asteroid at all. It is the dormant core of a giant comet.

At first sight, asteroids and comets seem to be very different kinds of beast. An asteroid is an insignificant speck of light, virtually invisible without a telescope, while a comet is the most spectacular of all sights in the sky. When a great comet appears, it puts to shame the planets, stars and Milky Way. The comet hangs in the sky with its tail stretching half-way from horizon to horizon, like a flaming sword.

To our ancestors, a comet was a portent of disaster. Shakespeare wrote in *Julius Caesar*: 'When beggars die, there are no comets seen; the heavens themselves blaze forth the death of princes'. In medieval and Renaissance times, every appearance of a comet caused a rash of broadsheets and pamphlets, describing the woes about to descend on mankind. In 1680, the astrologer John Hill published 'An allarm to Europe: By a Late Prodigious Comet . . . Together with some preceding and some succeeding Causes of its sad Effects to the East and North Eastern parts of the World'. Even in modern times, doom-mongers cash in when a comet appears. A pamphlet produced for Comet Kohoutek in 1973 proclaimed the forthcoming end of the world: 'Forty Days to Nineveh'.

A seventeenth-century engraving shows Halley's Comet over Jerusalem, at its appearance in AD 66. It was seen as a portent of doom; and indeed, four years later the city fell to the Romans.

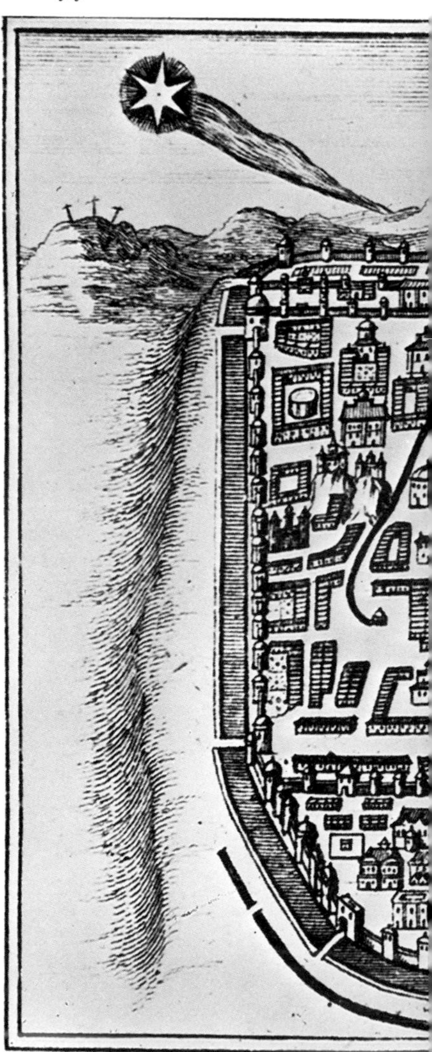

Astronomers are likely to get a similar treatment in the winter of 1985–6, when Halley's Comet makes its long-awaited appearance. Despite all the advance publicity, Halley's Comet will *not* appear as a brilliant sky sight, at least as seen from the northern hemisphere. When it is at its brightest, the comet will be behind the Sun, as viewed from the Earth and it will be lost in the Sun's glare. This is in fact the worst appearance of the comet in all of recorded history. But this comet does have an unparalleled importance in the history of comet research – and in its future.

Edmond Halley was a remarkable man, and a great scientist. He lived through a period of intense scientific activity in England, with the work of men like Robert Hooke, Christopher Wren (a scientist as well as architect), and Isaac Newton. Indeed, Halley would be better known as a scientist if he had not been overshadowed by Newton, probably the greatest scientist of all time. When Halley was four years old, these men founded the Royal Society, the world's leading scientific organisation; when he was nineteen, Charles II founded the Royal Observatory at Greenwich. Halley was then at Oxford; he left before getting his degree, and sailed to St Helena to make the first map of the southern stars. On returning to England, Halley made his greatest contribution to science, albeit an indirect one: he found that Newton had discovered the law that governs gravitation, but could not be bothered to publish it. Halley persuaded Newton to write up the theory of gravitation, and many other cornerstones of modern science, in a colossal work, the *Principia*; and Halley himself paid for the publication.

Halley's interests ranged widely. He invented a primitive kind of diving bell, and tested it in the English Channel; he worked out where Julius Caesar landed, on the south coast, near Rye. Halley sailed on three further journeys, in order to find the direction in which a compass needle points in different parts of the world – and on these voyages, he was the first civilian to be put in command of a naval ship. At an age when astronomers today must contemplate retirement, Halley was appointed Astronomer Royal. Although he was sixty-five, Halley decided that the pressing problem in astronomy at that time was fully to understand the motion of the Moon, a task that would take eighteen years to complete. Before his death, at the age of eighty-six, he had indeed finished these essential observations – and he died a happy and fulfilled man, with a glass of wine in his hand!

It is for Halley's study of comets, however, that he is most honoured. According to Newton's theory, comets must follow regular paths around the Sun, just as the planets

Edmond Halley was the first person to work out the orbits of comets. He found that comets seen in 1456, 1531, 1607 and 1682 were appearances of the same comet – now known as Halley's Comet.

do. The difference is that the planets' orbits are almost circular, while a comet's path is a long oval that takes it regularly from a point near the Sun (perihelion) where it is brightest, to a point beyond the known planets, where the comet is too faint to be seen. Halley had seen a bright comet in 1682. When he calculated its orbit, he found it was very similar to those of comets seen in 1607, 1531, and 1456. The interval between each appearance was seventy-six years (give or take a year or two). In 1705, Halley concluded that it was indeed the same comet, following a path that took it out to four times the distance of Saturn, and he correctly deduced that the slight changes to its seventy-six-year period were due to the gravity of Jupiter and Saturn. He predicted that the comet would return in 1758, long after he himself would be dead. And on Christmas night 1758, a German amateur astronomer spotted the comet as it returned to the Sun. The name of Halley was assured immortality.

With modern computers, we can easily trace the position of Halley's Comet back through recorded history. Don Yeomans, of NASA's Jet Propulsion Laboratory, keeps the most complete records and calculations on his computer. He can tell you where Halley's Comet was at any time back to 1404 BC. The first recorded sighting of Halley's Comet (rather than another 'bushy star') was in China in 240 BC. Somewhat surprisingly, the Chinese missed its next return; but in 1985, British astronomical historian Richard Stephenson found an account of the 164 BC appearance inscribed in Babylonian tablets.

In 837 AD, the comet barely brushed past the Earth, giving the most spectacular show on record, and the Chinese saw its head shining as brightly as Venus, following a tail that stretched over most of the sky. In 1066, the comet did indeed presage doom for England's King Harold. Halley's comet appeared in April, and six months later, Harold met his death at Hastings. The Norman conquerors naturally wove the comet into their commemorative Bayeux Tapestry.

So the comet keeps reappearing at an interval that is remarkably close to a human lifetime. We can reckon that each of us is likely to have one chance to see the comet; although a few people alive today were lucky enough to see the comet in 1910. (Among all the other Halley *memorabilia* on sale in the United States is a T-shirt proclaiming 'I'll see it again in 2061' – the T-shirt is, naturally, in small children's sizes only.)

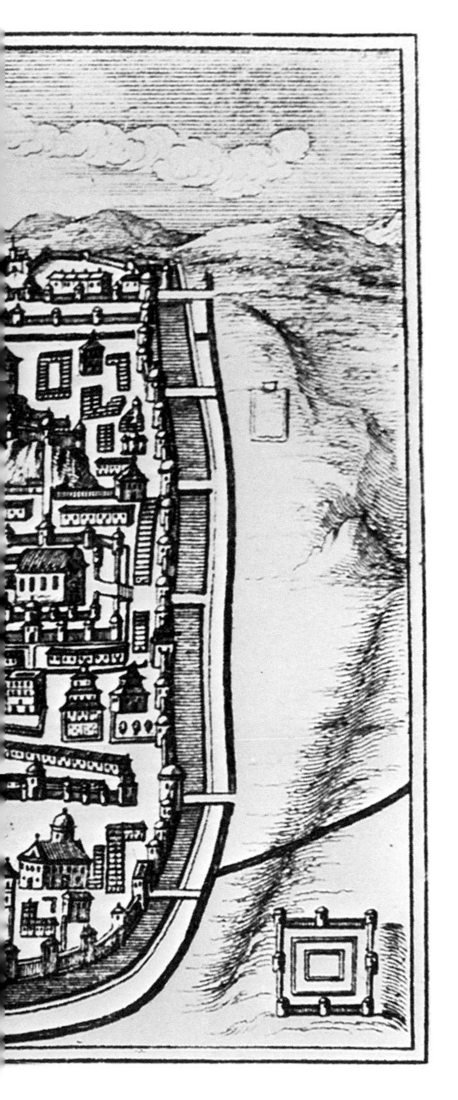

When Halley's Comet last appeared, astronomers still had little idea of what a comet was. The intervening decades have led to a generally-accepted picture, which astronomers want to test out on the unsuspecting Halley's Comet as it swings once more towards the Sun.

In the current view, a comet is essentially a dirty snowball, a few miles across (more technically known as a cometary nucleus). A large number of these objects live in a huge cloud that surrounds the Sun, well beyond the orbits of the planets. The Estonian astronomer Ernst Öpik first proposed this idea in the 1930s, and the Dutch Jan Oort confirmed it in 1950. The Oort cloud – as it's usually known – may stretch out half-way to the nearest star, so it marks the physical outer limit of our Solar System. The cloud may contain as many as a million million cometary nuclei, but despite this vast number, the total amount of matter in these small bodies amounts to only the mass of Mars.

Every so often, the gravitation of a passing star will shake one of these nuclei loose, and send it inwards on an orbit that will take it towards the Sun. Once it comes closer than Jupiter, the ices in the dirty snowball start to evaporate in the Sun's heat. They billow out into a huge glowing head (or coma), far larger than the planet Jupiter. Around the head is a still larger cloud of hydrogen gas, invisible except when viewed in ultraviolet light, that can grow to be several times larger than the Sun. The comet's head and surrounding hydrogen cloud become temporarily the largest objects in the Solar System; but their appearance is all show – they contain only a little matter, spread as a very tenuous gas.

As the comet comes in still closer to the Sun, it begins to sprout a tail – or, to be precise, two. One is an ethereal blue tail of gas that is swept away by the flowing 'wind' of electrically-charged particles coming from the Sun. The other is a yellowish tail, composed of small grains of dust that are carried away from the comet's nucleus with the evaporating gases; the tailshape is due to the pressure of sunlight pushing the dust

The Bayeux Tapestry, woven after the Norman Conquest of England, portrays King Harold hearing with terror of the comet in the sky. His courtiers, at left, 'marvel at the star', according to the inscription, while Harold thinks of the Norman invasion boats (bottom right). The comet's dread was justified: a few months later William defeated Harold at the Battle of Hastings.

away. In either case, the tail must always point away from the Sun. So we have the unexpected result that when a comet has rounded the Sun and heads away again, it is travelling tail-first.

After its brief encounter with the Sun, a comet may head right out to the Oort cloud again, or it may instead fall prey to the gravity of one of the giant planets. Jupiter or Saturn can easily swing a comet into a smaller path, so that it circles the Sun in only a few decades. It becomes what astronomers call a short period comet, because they can now predict the comet's regular returns to the vicinity of the Sun.

Larger particles of dust from the comet's nucleus are strewn along its orbit. These bands of dust are, in themselves, impossible to see; but they loop across interplanetary space like a minefield. When the Earth runs into a stream of comet dust, the tiny rock particles stream into the Earth's atmosphere by the million, and burn up as a shower of meteors. On an average night, you'll see a few shooting stars (meteors) every hour. These are stray pieces of dust which happen to run into the Earth's atmosphere. But when a meteor shower is on, you should see dozens, or even hundreds, of meteors every hour.

Each August, we see a rather dependable, and quite heavy, shower of meteors apparently coming from the constellation Perseus. This is the debris from a comet called Swift-Tuttle (after its two discoverers). Halley's Comet causes lighter showers in May and October. Occasionally, however, the Earth will plunge into a particularly dense stream of dust particles laid across space. When the Earth ploughed into the wake of comet Tempel-Tuttle, for example, in 1966, astronomers in the United States saw a veritable storm of meteors, at the rate of twenty every second!

But once a comet is in a short-period orbit, its fate is sealed. Each time it approaches the Sun and puts on its display of glowing gases, the comet loses some of its material. Like a winter snowdrift exposed to successive sunny days, the nucleus will eventually evaporate.

Last century, astronomers saw a comet shrivel away like this. In 1846, surprised astronomers saw Biela's Comet break in half; the two comets appeared together in 1852; but they have never been seen again. When the Earth passed through the one-time comet's orbit in 1877, however, the ashes of the dead comet rained down as a fantastic display of shooting stars.

On the other hand, a comet may lose its ices, but still retain its rocky constituents, which could end up as a small solid clinker. This rocky object would follow the same orbit about the Sun, typically an oval path taking it across the orbits of Mars and the Earth. Astronomers know of a couple of dozen asteroids in orbits like these. Traditionally, they have regarded these objects as refugees from the asteroid belt, flung towards the Sun at Jupiter's gravitational whim. But could they instead be the remnants of comets?

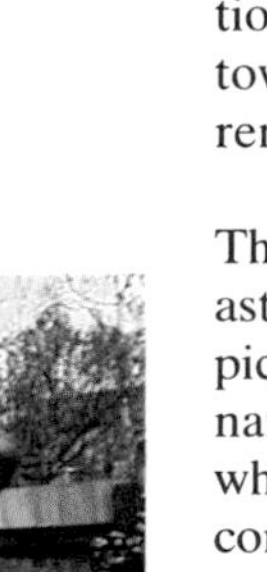

The authors celebrate the 1985–6 return of Halley's Comet with Comet Champagne. The comet's return always generates memorabilia, like T-shirts. At left is a model of the Giotto spaceprobe which will intercept the comet.

The orbiting infra-red telescope IRAS provided a vital clue. It could easily spot asteroids and comets, and a team from Leicester University analysed IRAS's results to pick out these fast-moving objects. In May 1983, John Davies found a comet that narrowly missed the Earth: it came closer than any object in 200 years, and people who caught the news bulletins saw the fuzzy comet-head move overhead through the constellation of the Plough. More exciting to astronomers, though, was a discovery by Davies' colleague Simon Green. He found an asteroid that pursued an elongated path, which took it closer to the Sun than any other known object. At its closest, this object, 1983TB, passes the Sun at only one-third Mercury's distance. Then other astronomers noticed a strange fact. 1983TB follows the same path as a stream of meteor particles, which causes a shower every December. No one has ever thought that an asteroid can strew dust particles along its track; so perhaps 1983TB is instead the remains of a comet.

If 1983TB is a missing link between comets and asteroids in the inner solar system, Kowal's 'mini-planet' Chiron may be its counterpart farther out. A comet nucleus coming in towards the Sun for the first time could be trapped in an orbit that keeps it out beyond Jupiter's orbit. Here, it is still so cold that its ices don't evaporate: the comet nucleus remains a small, solid body, looking just like an asteroid. So Chiron may well be a comet nucleus – but certainly a mammoth specimen. It's over twenty times the size of the nucleus of Halley's Comet, and if Chiron were ever pushed towards the Sun, it would produce the comet to end all comets.

But any comet is an excitement to the public. Even though this return will be disappointing, Halley's Comet is now spawning a million-dollar industry. In the United States, there are Halley T-shirts, bumper stickers, sports bags, glasses, balloons, jackets – and 'comet pills', to stave off any poisonous gases from the comet's tail. Telescope manufacturers are promoting Halley-scopes (most of them totally useless for observing something large and nebulous like a comet – you are better off with a pair of binoculars). In Britain, the Halley's Comet Society is producing Comet Champagne and Comet Whisky.

In November 1833, shooting stars fell 'like snowflakes' during a great meteor storm. Meteors are small particles of dust from a comet, burning up in the Earth's atmosphere. In this case, the dust is from Comet Tempel-Tuttle.

And if the view from Earth is a let-down, there should be grandstand shots from the first space probes to visit a comet. Because we know the comet's path so well, scientists in Europe, Russia and Japan could begin to plan their space probes even while the comet was so far off that no telescope could show it. Don Yeoman's computer kept day-by-day track of the distant and unseen comet as it headed back this way; and on 16 October 1982, astronomers at Palomar Observatory picked up the tiny, faint speck of light. It was right on course. Over seventy years had gone by since telescopes last showed the comet, but Yeomans had predicted its position along the path to within a few hours.

As we write, two Soviet 'Vega' probes are on their way to the comet, via Venus; the first of two Japanese craft, Sakigake, has taken off; and the Americans have sent a former Earth satellite to sniff the Sun's gases downwind of the comet. A second Japanese probe, Planet-A, should intercept the comet, too. A bevy of telescopes in orbit around the Earth will watch the comet, as will the camera on the Pioneer Venus Orbiter, its position near Venus giving it a unique view of the comet as it passes perihelion. But the most exciting pictures should come from the European mission, Giotto, which will plunge into the comet's heart.

The Italian painter Giotto di Bondone saw Halley's Comet in 1301; a few years later, he portrayed it as the star of Bethlehem in his 'Adoration of the Magi'. (In fact, the comet appeared too early, in 12 BC.) European space scientists have taken the painter's name for an audacious suicide mission. Using up-to-date information from Don Yeomans and from the Russian probes, the first to arrive at the comet, the European flight controllers will aim the sturdy little Giotto craft directly at the comet's nucleus. As it plunges down through the comet's coma, Giotto will run into the tiny grains of dust at a speed of 175000mph (280000kph): each speck of dust will hit the

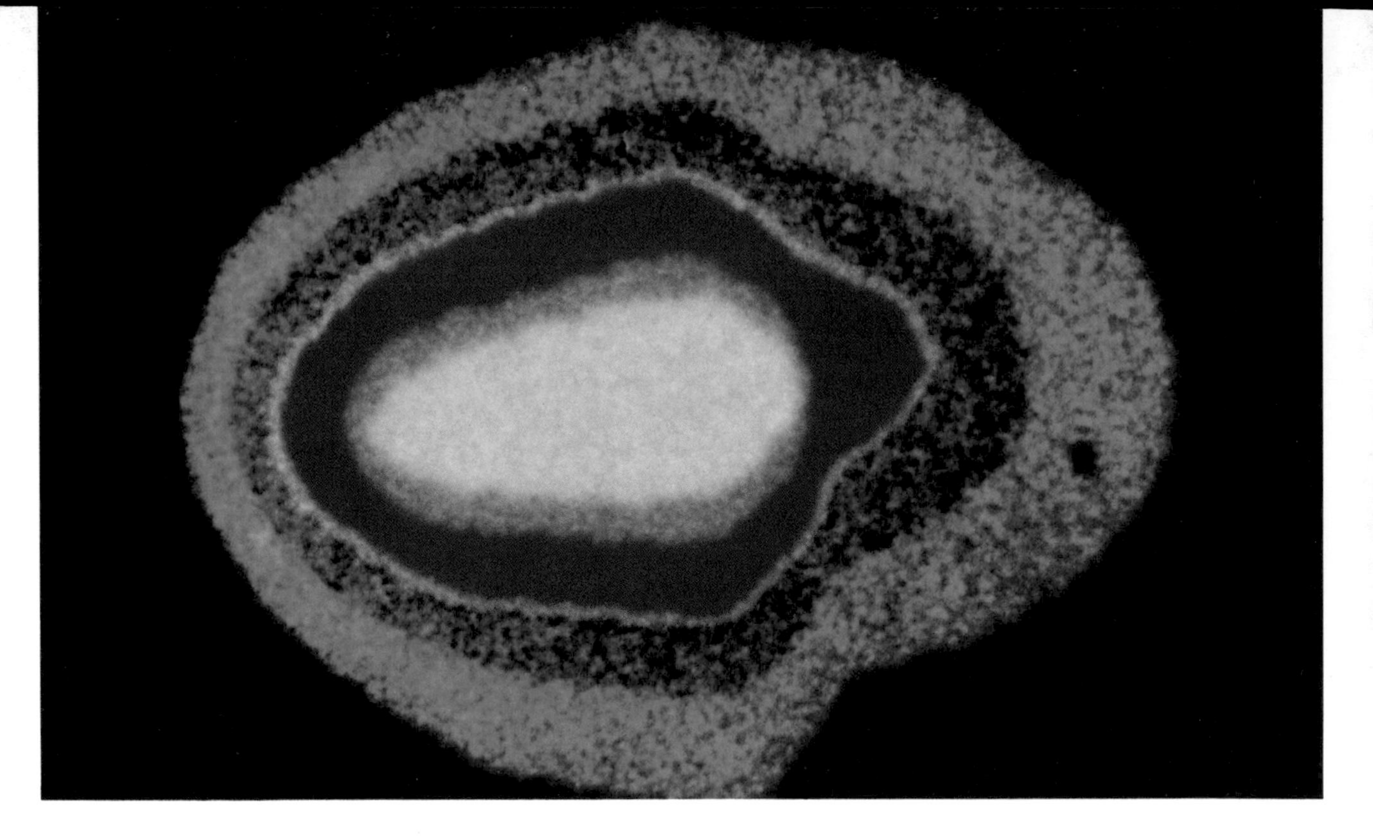

craft at almost 100 times the velocity of a rifle bullet. Giotto has an ingenious 'meteor barrier' to protect it from the worst of these impacts; even so, it may not last out the four-hour ordeal. All the while, however, its camera should be sending back more and more detailed pictures of the small world that lies at a comet's heart.

Although Halley's is the comet with popular appeal, scientists see it as only the beginning of a new phase of comet investigation. In particular, American astronomers feel they have lost out by not having their own Halley's Comet mission, and they are now planning a more audacious flight: a craft that will rendezvous with a comet, land on its surface, scoop up some of its ices, and return this sample from the deep freeze of the Solar System to the Earth. The mission may come off at about the same time as a proposed European probe to the asteroids. Both plans are symptomatic of a new perspective on the minor worlds of the Solar System. Astronomers no longer regard them as vermin or mere showy spectacles: asteroids and comets have much to teach us about the birth of the Solar System.

A 'heat image' of a comet which almost hit the Earth in 1983 displays in false colours the distribution of the dust in its head. This picture from the Infra-red Astronomical Satellite (IRAS) reveals a short dust tail that was too faint to be seen by ordinary telescopes. Two amateur astronomers also spotted this comet, and it is known by the joint name Comet IRAS-Araki-Alcock.

The Giotto spaceprobe plunges towards the heart of Halley's Comet, in this artist's impression of the kamikaze encounter in March 1986. The dish at the right sends the probe's pictures back to Earth; the camera is at the lower left part of Giotto. The front end (at left) of Giotto is built as a double layer to protect the craft from impacts with the comet's dust particles.

CHAPTER TWELVE

BIRTH OF THE PLANETS

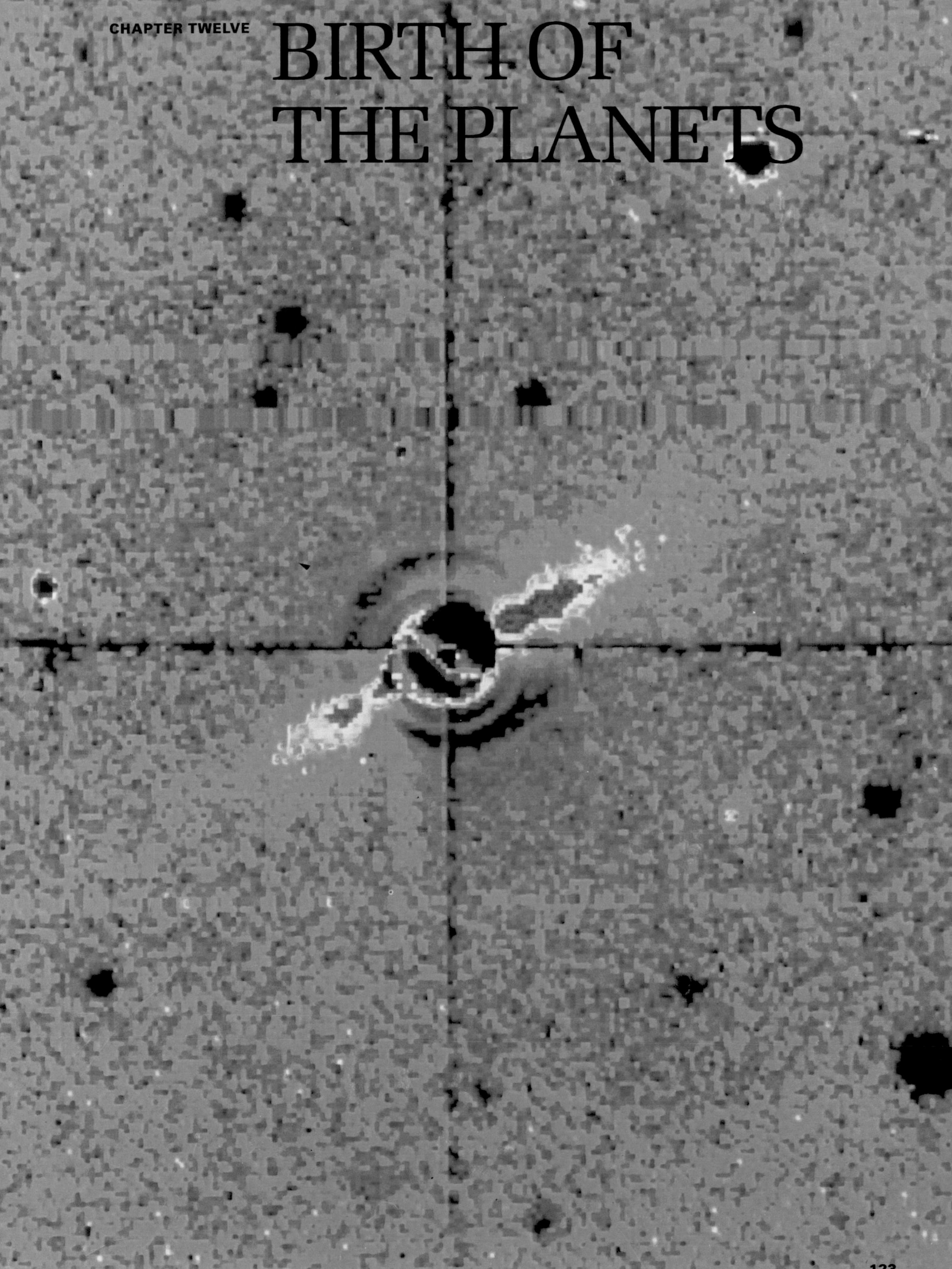

'In the beginning, God created the heavens and the Earth . . .' Like all theories proposed when Man believed the Earth to be the centre of the Universe, the account in the Book of Genesis links the birth of the Universe with the birth of our world. Indeed, in Greek mythology, the Earth came first. Before the moment of creation, there was only empty space, Chaos. The Earth goddess, Ge, sprang from Chaos, and then gave birth to the Heavens, personified by the god Uranus.

Once Greek philosophers began to look at the Universe in a logical way, they felt they could go to the other extreme. There was no need for a creation at all: the Universe has always existed. In the clockwork cosmos of Aristotle and Ptolemy, the planets, Sun and Moon follow their inexorable paths around the Earth, from the indefinite past to infinite future.

When the medieval scholars took over Ptolemy's Earth-centred Universe, they had to dovetail it with the Bible's account of Creation. This wasn't too difficult. When God made the Universe, he set up the various planets in their circular paths around the Earth. The detailed history given in the Bible also allowed Christian scholars to work out the date when the Universe came into existence. Archbishop Ussher of Armagh calculated that God created the Universe at nine o'clock in the morning of 23 October 4004 BC!

Even when Kepler and Galileo showed that the planets move around the Sun, rather than the Earth, there was no need to change this date. Sir Isaac Newton, who explained the planet's motions with his law of gravity, was a devout Christian who believed the Biblical account of Creation. According to Newton, God had acted in a simple and consistent way: when He made the material Universe, He also created 'laws' to govern it.

But a new breed of scholar was emerging, who wanted to explain the Universe on its own terms. Astronomers had shown that the stars are Suns in their own right, and it now became possible to look at the birth of stars (including the Sun and planets) as something distinct from the origin of the Universe as a whole. In the middle of the eighteenth century, astronomers had very little scientific evidence on how the planets were born, and the first ideas came from philosophers.

Among the earliest was the mystic philosopher Emanuel Swedenborg. He opined that the outer layers of the Sun condensed into a solid crust. A vortex around the Sun then pushed the crust outwards and broke it up, with the fragments becoming the planets. A French nobleman, le Comte de Buffon, had an alternative idea. A large comet fell into the Sun, and splashed out matter that condensed into the planets.

Another, much more soundly-based proposal, came from the German philosopher Immanuel Kant. His *General Theory of the Nature and Theory of the Heavens*, published in 1755, has an astonishingly modern flavour. Kant describes the Universe as infinite in size, and filled with 'systems and many stars' which are distant counterparts of the Milky Way galaxy. He bursts the bonds of the Biblical timescale, envisaging the passage of 'millions and whole myriads of millions of centuries'. During this period, stars condense from the primeval matter in swirling nebulae: the central part becomes a Sun, and the outer parts condense into planets.

The next step came from Sir William Herschel. After his discovery of Uranus in 1781, Herschel began to survey the stars with his giant telescope. In many cases, he could see a faint nebula surrounding a star, and he concluded that he was seeing a star being born from a cloud of surrounding matter. We now know that Herschel was wrong: such nebulae consist of gases expelled from dying stars. But his observations triggered a vital contribution to the theory of the planets' birth, a contribution that came from across the channel.

The great French physicist Pierre Simon Laplace had proved mathematically that the Solar System is stable – that the planets will always continue to follow paths similar to their present orbits – he began to investigate the origins of the planets. Independently of Kant, Laplace came up with a 'nebular theory'; and when he heard of Herschel's observations, he began more detailed calculations. Laplace envisaged that the nebula would shrink towards the Sun, and every so often the outer parts would be shed as a gaseous ring, that would condense to form a planet.

Previous page: *This historic picture is the first view of planets forming around another star. In this false colour-view, the star itself – Beta Pictoris – is hidden by a mask at the centre of the picture. The yellow and red 'wings' are a disc of dust and gas that we are seeing edge-on. This material is almost certainly condensing into planets, in the same way that a disc of matter condensed into the planets of our Solar System almost 5 billion years ago.*

A German engraving of 1744 expresses the idea that beyond our Solar System there are similar families of planets going around the other stars. The outermost planet of the Sun's family known then was Saturn, shown at left with four moons; the oval track around the Sun is the orbit of a comet.

Early in the twentieth century, the nebular theory went out of fashion. Several scientists, including Sir James Jeans in England, found that it presented problems. The nebula would not throw off rings, and even if they did, the gas would evaporate into space, rather than condense into planets. Most damning was the calculation that most of the rotational momentum in the Solar System would end up in the Sun. In fact, the motion of the planets carries a much larger quantity of momentum than is bound up in the Sun's rotation.

Jeans and his colleagues proposed a completely different idea: the 'tidal theory'. Some time well after the Sun itself was born, another star happened to whizz past, close to its surface. The star's gravity pulled a cigar-shaped filament of gas from the Sun (in an enormously exaggerated version of the tides that the Moon raises on the Earth). This gas condensed into the planets, with the thickest region in the centre forming the giant planets Jupiter and Saturn. The motion of the star imbued the filament with a lot more momentum than the Sun contains. Although this theory was popular for a while, it was really a non-starter, because the hot gases torn from the Sun would certainly not condense into planets, but promptly boil off into space. In recent years, astronomers have been able to put the last nail in the tidal theory's coffin, as they have found that the planets contain proportions of some rare chemical elements (like lithium) which are similar to the proportions in the gases of interstellar space, but very different from those in the Sun (where nuclear reactions destroy lithium).

Both theories seemed to have one insurmountable problem: the gas near the Sun would not condense into planets. But even as Sir James Jeans was working on his tidal theory, Robert Trumpler at California's Lick Observatory made a discovery that solved this problem – but only for the nebular theory. Trumpler found that the gas between the stars is mixed in with small amounts of solid dust, in microscopic grains. In modern versions of the nebular theory, it is the solid grains collecting which initiate the birth of the planets, rather than the gas condensing. The nebular theory's problem with momentum has now also disappeared. In the original rotating disc, the Sun could have transferred momentum to the surrounding gases in at least two ways – either by its magnetic field, or by a gravitational pull, if the early Sun was somewhat sausage-shaped. In fact, there's no need even for this. In the past few years, astronomers have discovered that very young stars have a violent 'wind' of gases blowing from their surfaces. When the Sun was young, this wind could have blown out between the planets, and carried away the Sun's excess rotational momentum.

So most astronomers now reckon that an updated version of the nebular theory is the best bet. It fits in with our observations of the regions of space where stars are being born now; and it fits in with our analysis of the oldest rocks in the Solar System.

Astronomers have now found that stars are formed in dense clouds in space, unfortunately so packed with interstellar dust that these dark grains totally obscure what's happening within. But even if light cannot penetrate the densely-packed dust, the longer wavelengths of infra-red and radio waves can get through. Telescopes designed to detect these radiations can show us where stars are being born, and how the matter around these stars is moving. It turns out that an infant star is generally swaddled in a rotating disk, thick with dust and gas, just as the nebular theory predicts.

We have some direct evidence of the birth of our own Sun and planetary system from very old rocks. These are not rocks from the Earth, which have been much-altered by geological activity; nor the Moon-rocks, gathered at such expense, because these have been battered by the impact of great meteorites. The 'birth certificate' of the Solar System – the oldest rocks we know – consists of matter from the darker minor planets of the asteroid belt, which can be examined in the form of meteorites that land on the Earth.

On the night of 8 February 1969, people throughout Mexico awoke in fright to see a brilliant blue-white light crossing the sky. The inhabitants of Pueblito de Allende were even more terrified. As it came towards their village, the fireball broke in two, and the twin lights then exploded, showering the area with rocks. This brilliant object was a meteorite, and investigators eventually found two tonnes of extraterrestrial rock strewn around the village. The Allende meteorite could hardly have fallen at a better time, or a better place. American scientists had just built a laboratory for analysing the Moon-rocks, to be brought back by Apollo astronauts later that year. They took parts of the Allende stone across the border, and subjected it to intimate scrutiny.

The meteorite turned out to be even more of a godsend than scientists had expected. It was of a rather rare type, a stony meteorite whose material was darkened with carbon. Astronomers have found only forty of these carbonaceous chondrites among the thousands of meteorites that are known – and Allende was one of the biggest. According to previous studies, carbonaceous chondrites are the oldest rocks known, fitting in with the idea that they have been chipped away from small asteroids that have remained unchanged since the birth of the Solar System. Scientists can measure the age of a rock by looking at the concentrations of radioactive material in them: these atoms break up gradually as time goes by. The Allende meteorite is 4550 million years old – and that must be the date when the Solar System was born.

We can now place the formation of our planetary system within the 'history' and 'geography' of the Universe as a whole. Astronomical measurements tell us that the cosmos began about fifteen billion years ago, in an all-encompassing explosion called the Big Bang. As the gases spread out, they soon cooled and condensed into galaxies – each a collection of a billion or more stars. In the spiral-shaped galaxies, some gas was left over, and over the ensuing eons it gradually condensed to form stars. At the same time, dying stars pumped gases back into space. The occasional star would blow itself apart in a spectacular explosion, called a supernova, temporarily outshining 10 billion Suns. The shock wave from a supernova can squeeze the surrounding gases in space, and so trigger the birth of whole new generation of stars.

When our Galaxy was already ten billion years old, something happened in its outer reaches. This event was of little importance to the Galaxy as a whole; but its effects proved to be profoundly important to us. A supernova's remains cannoned into a gas cloud in space, compressing it so much that its central regions began to condense into stars. One of these stars was our Sun.

As this fragment of whirling gas and dust shrank in size, it flattened into a disc. In the middle, it was most compressed and this squeezing made it heat up. The central hub shrank more and more, until its temperature reached ten million degrees. Now the atoms of hydrogen gas were colliding so fast that they began to react together, in a process of nuclear fusion. The reactions created an enormous amount of energy, power that flowed outwards and prevented the hub from shrinking further. As it steadied in size, the gas sphere blew off much of its outer layers of gas, in the process probably ridding itself of large amounts of rotational momentum. Then it settled down as a steadily burning star – our star, the Sun.

Meanwhile, the surrounding disc of matter took the first steps towards becoming a system of planets. In the outer regions, the temperature stayed close to the profound cold of interstellar space. Here the grains of solid matter consisted largely of ice. Closer into the newly-born Sun, the temperature was considerably higher. Within a distance of 300 million miles (500 million km) of the Sun, the heat was fierce enough to evaporate away the ices, and the only solid grains were particles of rock. In both parts of the system, the tiny grains began to stick together and build up larger blocks of materials – hunks of warm rock near to the Sun, and frozen snowballs further out.

At the moment, scientists cannot calculate precisely how these blocks grew in size. That's why planetary astronomers want to scrutinize the rings of the giant planets. Saturn's rings, in particular, seem to be a mêlée of icy fragments forming and reforming: a microcosm of the early Solar System, preserved for our inspection because Saturn's gravity breaks up the larger icy chunks as fast as they form. We can see that the particles in Saturn's rings, small as they are, exert a gravitational pull on each on other that marshals them into innumerable ringlets. The larger chunks in the Sun's disc would have caused an even stronger effect, that clumped them together into larger bodies, each about a mile across, called planetesimals.

But it is still a long way from millions and millions of planetesimals, each a separate 'mini-planet', to a handful of large worlds orbiting the Sun. In the 1960s, the Soviet scientist Victor Safronov made detailed calculations about what happens when two planetesimals pass close to one another. He found that the bigger planetesimals would sweep up the smaller ones; and as they grew yet larger in the process, the big planetesimals would get even more efficient at catching the smaller ones. As a result, the rocky planetesimals near to the Sun were swept up into just four worlds – Mercury, Venus, Earth and Mars – or possibly five, if the Moon began life as an independent world. At every stage, the infalling planetesimals must have blasted out craters on the

Stars and planets are being born in this colourful riot of gases in the constellations Scorpius and Ophiuchus. Star birth actually occurs in the dark clouds which here obscure the light from stars and nebulae behind. Once they begin to shine, the young stars light up the surrounding gas clouds as multicoloured nebulae.

surface of the forming planet, and the pockmarks we see today are only the final signs of a battering that has lasted since the first couple of planetesimals came together in the conception of the planet.

Beyond the orbit of Mars, there is a region where the gravity of the young Jupiter prevented the planetesimals from making a single planet. Instead, they grew into the thousands of small worlds of the asteroid belt. The fragments of asteroid that land on Earth as meteorites tell us not only the age of the Solar System. A more detailed study of their radioactive elements can reveal how long it took for the original dust to form into asteroids – which must be roughly the time that it took for the Solar System to form. The answer comes to about 100 million years. Although this is a long period in human terms, it is only an instant when compared to the age of the Solar System: we can say that the Solar System began 4550 million years ago, and the planets were essentially complete by 4450 million years ago. (The Moon-rocks reveal that the mopping-up operation on the remaining planetesimals took another few hundred million years.)

Something rather different must have happened in the icy outer reaches of the Solar System. The bulk of Jupiter is made up of hydrogen and helium gases, which could never have frozen into solid planetesimals. For a long time, most astronomers thought Jupiter had condensed straight out of the gases of the original disc. The solid icy particles trapped inside the gas globe would have sunk to the centre to form Jupiter's core. But now it seems that this was exactly the wrong way round: the dense core came first.

The Japanese astronomer Chushiro Hayashi was the pioneer of the new theory. He assumed that the outer parts of the Solar System were filled with icy planetesimals, just as there were rocky planetesimals nearer to the Sun. The icy chunks accumulated, and grew into four worlds, each about fifteen times heavier than the Earth. According to Hayashi's team, something remarkable then happened. The whole region around these planets was surrounded by the hydrogen and helium gases of the great disc, and each planet's gravity had attracted the neighbouring gases towards it. When the central icy world grew to a weight of fifteen Earth-masses, it acquired a substantial atmosphere of hydrogen and helium, which settled around the planet, giving it extra gravity to pull in more gases – which in turn increased its gravitational pull still further. In a very short time, the planet raked in all the gas near its orbit. Only the amount of gas available limited its size.

In this picture, Jupiter started off as an icy world of fifteen Earth-masses. Located in the densest region of the gas disc, it managed to attract twenty times more matter than it originally contained. Saturn, farther out, had a skimpier supply to draw on, and it grew by a smaller amount. Distant Uranus and Neptune were in a region with very little gas, and they acquired only a thin atmosphere of hydrogen and helium which added little to their bulk.

The small bodies of the outer Solar System – Pluto, Chiron and the moons of the outer planets – are icy planetesimals which managed to avoid being gobbled up by the larger worlds. The huge swarm of comet nuclei, in the extensive Oort cloud, may have a similar origin; but here there is still some mystery. Way out in the region of the Oort cloud, the original nebular disc was so tenuous it seems unlikely that the microscopic ice grains could have met up and clumped together. Some astronomers have suggested, instead, that the comet nuclei were planetesimals that grew up in the shadow of the giant planets. The four great worlds did not sweep them up, but instead flung them out into enormous orbits that took them far from the Sun.

Our studies of starbirth, the planets and meteorites now firmly point to the birth of the Solar System from a nebula of gas and dust swirling around the young Sun. This disc was apparently a natural part of the formation of the Sun, which is a very average star. So, do other stars also have planets?

This is one of the most exciting questions in astronomy today. For years, there has been argument; now, we are on the verge of getting a definitive answer. The insurmountable problem has always been that we cannot hope to see planets of other stars directly. Since a planet has no light of its own, but only reflects the light of its parent star, the planet will always be far, far fainter than the star. The star's light is so dazzling that we cannot discern a faint planet nearby: it's like looking at a searchlight that is directed towards you, and trying to make out a glowworm on the grass below.

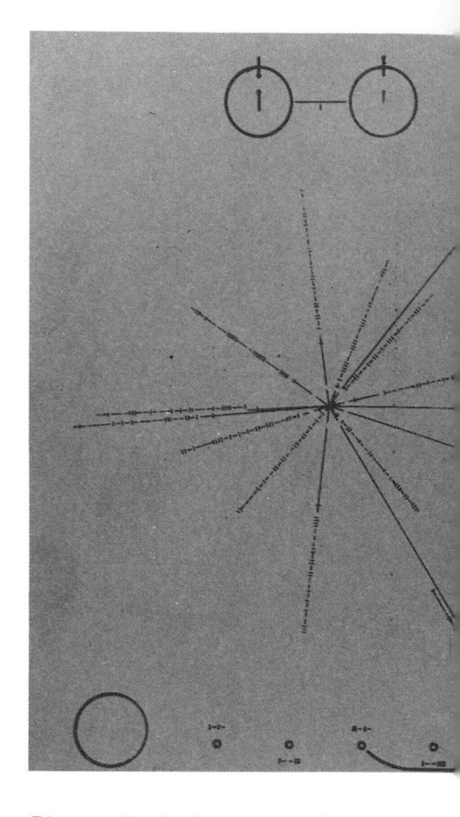

Pioneer 10, the first spacecraft to leave the Solar System, carries this plaque to identify where it came from. An alien who may come across the probe is given a starmap (left) which shows where the Sun lies, relative to the radio-emitting pulsars in the Galaxy. At the bottom is a diagram of the Solar System, with Pioneer starting from Earth and swinging past Jupiter. Unless the alien is used to the conventions of line-drawing, it

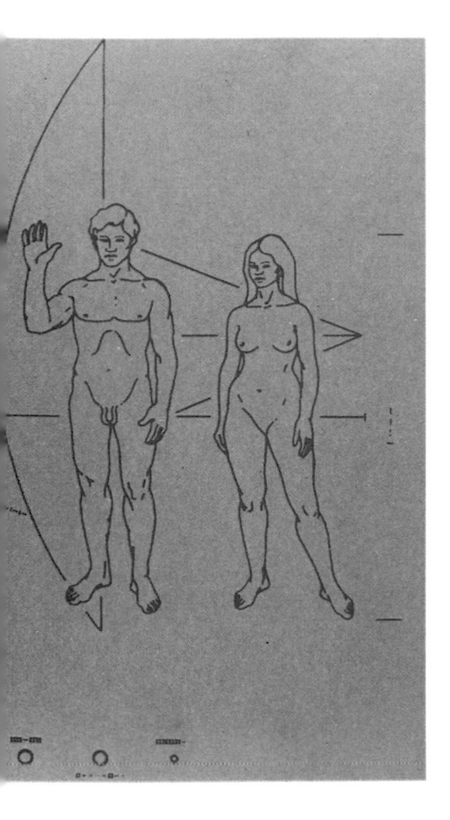

may find great difficulty in working out what the complex curves on the right are meant to represent!

A more detailed message to the stars was loaded on to the Voyager 1 and 2 spaceprobes. A record (left) carried the 'sounds of Earth', including an erupting volcano, a baby crying, music from Bach to Chuck Berry, and greetings in all the Earth's main languages – including whale-song.

But there is an indirect way. When we say that a planet is orbiting a star, it's actually a simplification. Both the planet and the star are moving in orbit about their balance point, or centre of gravity. This imaginary point is always closer to the heavier of the two bodies; and with a planet and a star the centre of gravity is so close to the star that we usually refer to the star as the centre of the planet's orbit. But in fact the star is moving, too. Jupiter's gravity actually swings the Sun around a point that is outside the Sun's surface. So if we watch a distant star, and find that it is swinging around in space, we can be sure that the star is not alone.

For the past fifty years, Peter van de Kamp has been keeping a close watch on several stars that are near to the Sun. Some of the stars move in dead straight lines through space, and must be solitary. But van de Kamp has discovered that others follow a slightly wiggly path: these stars must have unseen companions that swing them from side to side as they travel along. The celestial balance shows that in most cases the faint companion is too heavy to be a planet, and must be a very dim star. There is one famous exception, however: Barnard's Star.

This faint red star lies only six light years from the Sun, and van de Kamp's painstaking observations show that it has a small 'wiggle' to its motion through space – so small that it must be the effect of a planet rather than a companion star. He goes so far as to analyse the star's motion as the result of two planets, very similar to Jupiter and Saturn. Unfortunately, other astronomers have failed to confirm van de Kamp's measurements on Barnard's Star.

One of America's leading experts on star and planet motions is Bob Harrington, who was involved in the discovery of Pluto's moon and who is currently searching for a tenth planet in the Solar System. Harrington is now in charge of a project that has led to the first definite evidence for planets of another star. Using the US Naval Observatory's purpose-built telescope at Flagstaff, Arizona, he has been studying the motions of almost a thousand stars. Of these, about a dozen seem to wiggle slightly on their journeys through space. For ten years, Harrington has been keeping a special check on two faint stars relatively near to the Sun, VB8 and VB10. They are two of the dimmest stars known: if VB10 were put at the centre of the Solar System in place of the Sun, it would give scarcely more illumination than the full Moon. And both show signs of a definite wiggle. Harrington estimated that each star had a companion a few times heavier than Jupiter – but he isn't inclined to be too definite about something that no one could see.

In the meantime, the new science of infra-red astronomy has been fast maturing, and it promises to be the ideal tool for planet seekers. A star like the Sun looks relatively faint when 'seen' by its output of infra-red rays; on the other hand, a planet is 'brighter' in infra-red than when seen in ordinary light. To our eyes, a 'Jupiter' going round another 'Sun' would look a billion times fainter; but an infra-red sensor would see the planet only 10 000 times fainter, and so it would be less dazzled by the star's glare. What astronomers needed was an 'infra-red eye'.

This is where Frank Low came in. Brought up as a physicist, he made the world's first sensitive detector for infra-red – and turned to astronomy to demonstrate his device. In the early 1970s, Don McCarthy came to work with Low, and they made a special infra-red sensor that they could put on a telescope, and 'see' the stars in a double star pair as separate images. Their equipment worked better than Low and McCarthy had expected. When they looked at one well-known double star, they found a dim third star that just wasn't visible with an optical telescope.

Don McCarthy took a list of stars that van de Kamp, Harrington and other astronomers had suspected to have unseen companions. He quickly saw that some did have faint stars next to them. When Low and McCarthy looked at the star VB8, from Harrington's list, they got more excited. Yes, there was a companion. But it was too dim, and too cool, to be a star. Putting together Harrington's information on how the companion waltzes with VB8 itself, and their infra-red picture, Low and McCarthy deduced that they were seeing a planet, ten times heavier than Jupiter – the first planet of another star.

But this world seems to be rather different from the planets we know. It is so much heavier than Jupiter that it's quite hot, with a temperature over 1000°C. In fact, this new object, called VB8B, poses the question 'when is a planet not a planet?'. In the

Solar System, we can easily draw the distinction between a planet and a star – in the shape of the Sun. But the difference is more in quantity than quality. The heaviest planet, Jupiter, is made up of the same kind of gases as the Sun. If we added more of the same matter to Jupiter, its centre would get hotter and hotter, until the temperature was high enough for nuclear reactions to start. Then our 'super-Jupiter' would shine as a star. This dividing line between planets and stars comes with an object about 80 times heavier than Jupiter.

So we could call VB8B a planet – but most astronomers feel it's so different from the planets of our Solar System that it should have a new kind of name. It is a 'brown dwarf'. Shiv Kumar, who first calculated the dividing line between planets and stars, dreamt up the name long before Low and McCarthy found the first brown dwarf in the sky, and it has become something of a catchword among astronomers, along with the other colourful phrases – 'red giants', 'white dwarfs' and 'black holes'.

A brown dwarf is much hotter than a planet, because it is still shrinking from its birth in an extended cloud of gas. Frank Low points out that this description applies to Jupiter, too: as a result of its contraction, Jupiter produces as much heat in its core as it receives from the Sun. We can then argue that the Solar System doesn't consist of one star and its nine planets; but one star, eight planets, and one brown dwarf!

Whatever we call VB8B, the discovery is a breakthrough in the search for planets of other stars. McCarthy is continuing to look for companions to other stars – particularly Harrington's other candidate, VB10, which may have a lightweight companion. Low concedes that a telescope on the ground would not be able to detect a Jupiter going around another star – but it should be easy with an orbiting telescope, above the Earth's perturbing atmosphere. To settle finally the question of the planets of Barnard's Star 'would be a five minute job'. Even so, it wouldn't be possible directly to see a planet as small as the Earth.

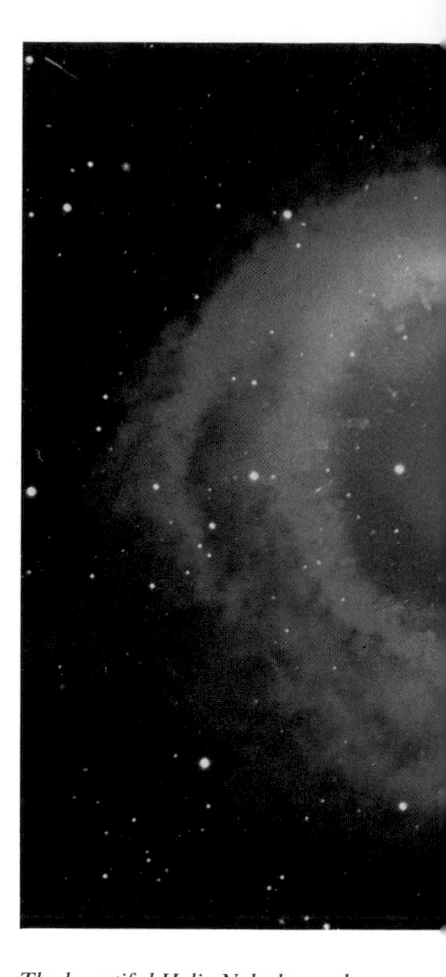

The beautiful Helix Nebula marks the death of a star, and of its accompanying planets. Astronomers call such dying stars 'planetary nebulae', because the shell of gas looks like the dim disc of a distant planet, when seen in a small telescope. But this old star must have destroyed any planets close by, as it began to swell in its death throes. The Sun will eventually die as a planetary nebula, in about 5 billion years' time. But we can be confident that, by then, mankind will be able to cross the vastness of interstellar space, and resettle on the planets circling other, younger stars – stars and planets that have yet to be born.

NASA scientist David Black hopes to pin down other planets with a special optical telescope in space. Above the 'twinkling' of the atmosphere, it could measure the star position with a precision we cannot achieve on the ground, so allowing Black to measure the wiggles in the paths of many nearby stars, this telescope could measure the effect not just of 'Jupiters' or 'Saturns', but also of planets like Uranus and Neptune. Black's telescope will be able to show if our system is typical; and it may even be able to reveal the gravitational pull of an 'Earth' going around another star.

Black's telescope is a proposal for the future. Bob Harrington thinks we may be able to see larger planets – brown dwarfs, certainly – with a telescope that's just about to be launched. The Space Telescope is a fairly ordinary optical telescope, but at an extraordinary site; in orbit, where it is free from the twinkling, blurring and stray light caused by the Earth's atmosphere. Meanwhile, other astronomers, including Bill Cochran at Texas's McDonald Observatory, are trying a new tack with existing telescopes on the ground. For a century, astronomers have measured the speed of stars towards us or away from us, by analysing their light in detail. Now we may be able to detect the tiny change in the star's speed as it swings around the balance point with a planet.

There's a feeling of excitement among astronomers at the moment. We now know that a star, VB8, has a companion that is 'planet-like' – as Frank Low puts it. We have the equipment to find 'Jupiters' going around other stars. And the momentum is there to construct instruments that could prove there are planets like ours elsewhere in the Universe. Frank Low admits, 'in my lifetime, it won't be possible to build an infra-red detector to find Earths around other stars – which surely must exist. But we could do so by combining techniques.' David Black comments that the advent of the new techniques 'will make it possible to answer soon one of the oldest and most significant questions raised by mankind'.

Even without seeing the planets of other stars, astronomers are now feeling confident that planetary systems must be common in our Galaxy. The evidence has come from the pioneering infra-red telescope, IRAS. This supercooled eye on space was launched in January 1983, and from its orbit it scanned the sky for sources of heat in the Universe. IRAS shot to fame with its new and totally unexpected discoveries. Of these, the most exciting was the birth of a planetary system around Vega.

American researchers George Aumann and Fred Gillett were sitting back-to-back in an office in Britain's Rutherford Appleton Laboratory, on what started as a perfectly normal day for analysing the information that was streaming down from the satellite above. They had instructed IRAS to look at some bright stars, to check its telescope was pointing in the right direction. Aumann had insisted that they include Vega, the fifth brightest star in the sky, even though it is so hot that little of its radiation comes out at the wavelengths that IRAS could 'see'. That's why Gillett was puzzled. The computer printout showed that IRAS was seeing Vega much brighter than it should be. He called Aumann over, and they tried to work out what had gone wrong. When they found that the satellite appeared to be working perfectly, Aumann realized there was only one answer. The hot star Vega must be surrounded by a cooler ring of dust, and the dust was producing the infra-red rays. He peered in detail at IRAS's sweeps across Vega, and yes, indeed, the telescope was seeing something larger than the star. They observed and re-observed Vega for a month, and there was no doubt. The star is surrounded by a dust ring, twice the size of the Solar System.

Vega has exactly the kind of surrounding dusty disc that we think formed into planets around the Sun. Add to that the fact that Vega is a comparatively young star, and Aumann and Gillett had little doubt that they were seeing the birth of a planetary system. Alerted by Vega, they came across almost twenty more examples. The star Fomalhaut, in the constellation of the Southern Fish, is another bright star with a dusty disc, an almost identical twin of Vega. Aumann took the statistics of stars with dust discs and without and related them to the ages of the stars, to come up with some fascinating answers. First, the discs seem to last for about 100 million years, exactly the period that it took for the Solar System to form, according to the evidence from meteorites. And secondly, the dust discs are common enough that almost every star might have been born with a dusty halo. The statistics are telling us that when IRAS finds a star without dust, it is a star older than 100 million years. What has happened to its dust? It must have condensed into planets – which IRAS could not pick up individually. So there's persuasive evidence that many stars in the sky have a retinue of planets.

In Aumann and Gillett's list, one star stood out as a brilliant source of infra-red. To optical astronomers, it was a fairly insignificant star in the southern sky, called Beta Pictoris. It so happened that Brad Smith and Rich Terrile, two astronomers interested in the Solar System's outer planets, were at the Los Campanas Observatory in Chile, with a special camera that could record faint details right next to bright objects. This camera proved itself by taking the first picture of the light reflected from Uranus's dark rings. Hearing of the IRAS results, Smith and Terrile pointed the telescope at Beta Pictoris – and they found that the star is flanked by two faint streaks of light extending away from it in opposite directions, looking rather like Saturn's rings seen almost edge-on. Smith and Terrile had photographed directly the disc of dust and gas in a forming planetary system, seen edge-on from the Earth. To date, this is the only picture of a planetary system – albeit an embryonic one – beyond the Solar System.

The mid-1980s are seeing an upsurge in the belief that there really are planetary systems around other stars. When we can see them and investigate them in more detail, the results will be immensely important for the study of the worlds of the Solar System. So far, we have only one system of planets to photograph, prod and sniff at. Science needs more than that: where would botany be if we had only one small garden to study; or geology if we had access to just one quarry?

But there's another consideration that is not far from the thoughts of even the most hardbitten scientist. The planetary systems of other stars may be amazingly diverse and varied, but the basic laws must still hold true. If there is a medium sized rocky planet at just the right distance from its star, liquid water should run over its surface and collect in its hollows. The common element carbon should be abundant on this world. The chemical processes that make up carbon molecules in places as diverse as dark asteroids, the clouds of Titan, and the gas clouds of interstellar space, should bond together the elements on this watery world, as surely as they did on the primeval Earth.

And the processes of organic change, and the survival of the most persevering species should ensure that life evolves on this planet, and evolves to become intelligent. With the first awakening of intelligence, planet X around star HD 963251, say, will seem to be the centre and hub of the Universe. Later, the true scale and grandeur of the Universe will fall into place. And as they begin to explore the planets in their own little corner of the Cosmos, these beings too will look up to the sky – perhaps towards an insignificant yellow star that we know so well – and begin to wonder . . .

APPENDIX ONE

PLANET SPOTTING

Although the brilliant planets Venus and Jupiter draw immediate attention to themselves in the sky, it's sometimes more difficult at a glance to tell the fainter planets from stars. One way is to check if the suspect moves from week to week; but most of us require an answer more quickly than that! A good test is to see whether the object twinkles or not. Unless a planet is very low down in the sky, and we are seeing it through the densest layers of our atmosphere, planets do not twinkle (as explained in Chapter 2). And so anything which shines with a steady glow is probably a planet, and not a star.

If you want help in locating the planets, or want to know which planet you've spotted, there are several places you can turn for help. A number of leading newspapers publish a monthly guide to the sky; and there are also yearbooks of astronomy published annually with a guide to the stars and planets month-by-month. We've also listed here (next two pages) where you'll see the planets at any time up to 1990, with

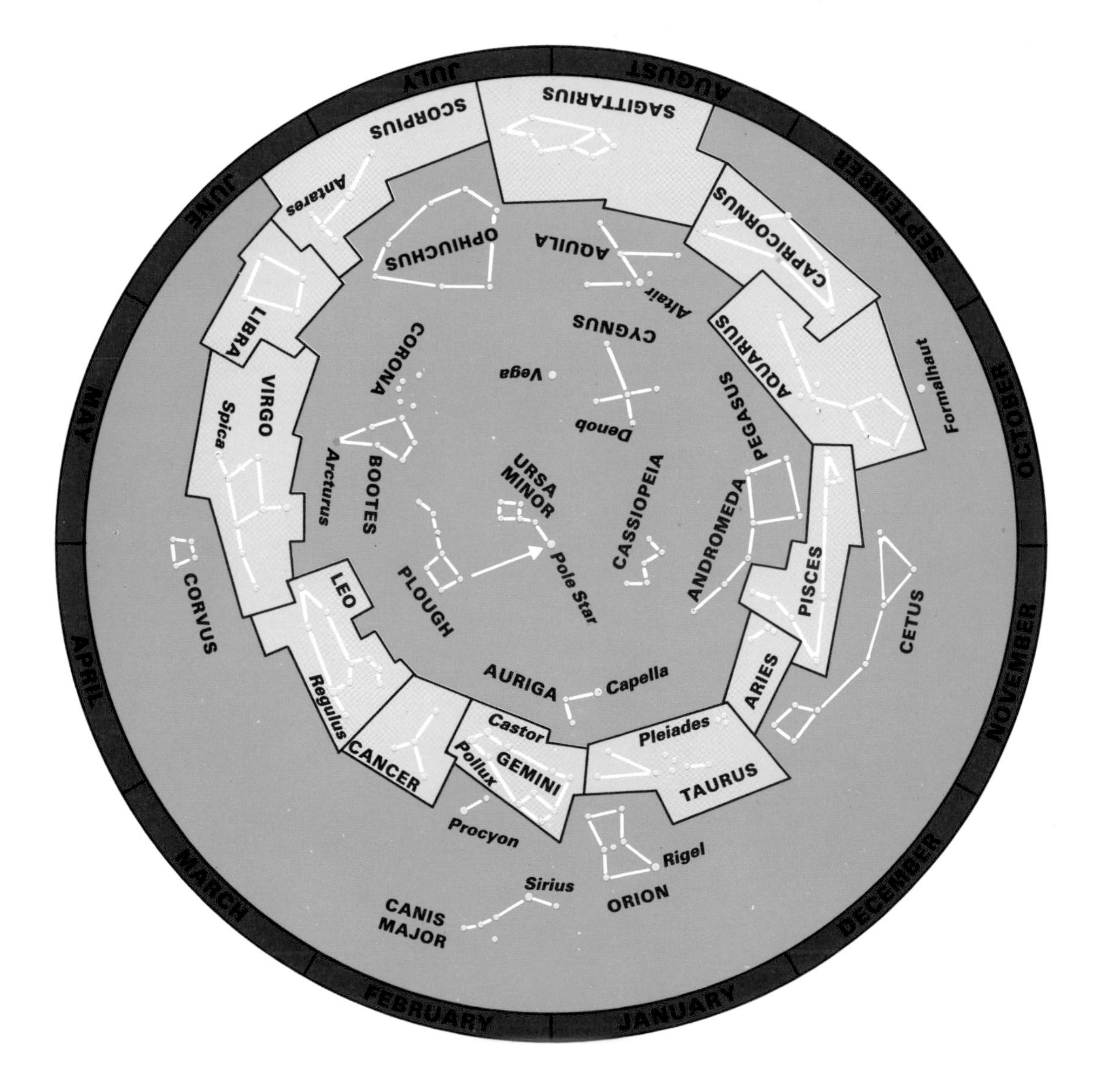

charts to help you locate them. To do this, you'll need to identify some of the star-patterns (constellations) in the sky – and the two maps on this page are a simple guide. Use the top map if you live in the northern hemisphere; the lower map if you are south of the equator.

Let's take the northern map as an example. It shows all the stars you can see from, say, Europe or North America, during the course of a year. But on any particular night, only two-thirds of these constellations will be above the horizon. To spot the stars and constellations, first locate the Pole Star. (If you don't know where it is, find the Plough – or Big Dipper – and follow the end two stars, as shown.) Stand with your back to the Pole Star, and hold this page vertically in front of you. Rotate the book until the current month appears at the bottom of the star chart. The stars in the lower part of the map should now match up to the stars you see in front of you, from the horizon up to the overhead point. (Be careful with the scale – the constellations in the sky will look much bigger than they do on the page.)

The planets always lie in one of the constellations that are coloured here (the zodiacal constellations). Identify these star patterns by some of their brighter stars (bearing in mind that an unexpected bright 'star' might be a planet), and then turn to the more detailed charts over the page.

If you live in the southern hemisphere, you don't have a bright pole star. If you know the compass directions, stand with your back to north. Otherwise, use the Southern Cross as your pointer: the position corresponding to the 'Pole Star' is half-way between the Cross and the bright star Achernar. Once you've worked out which direction to face, follow the above instructions, using the lower star chart.

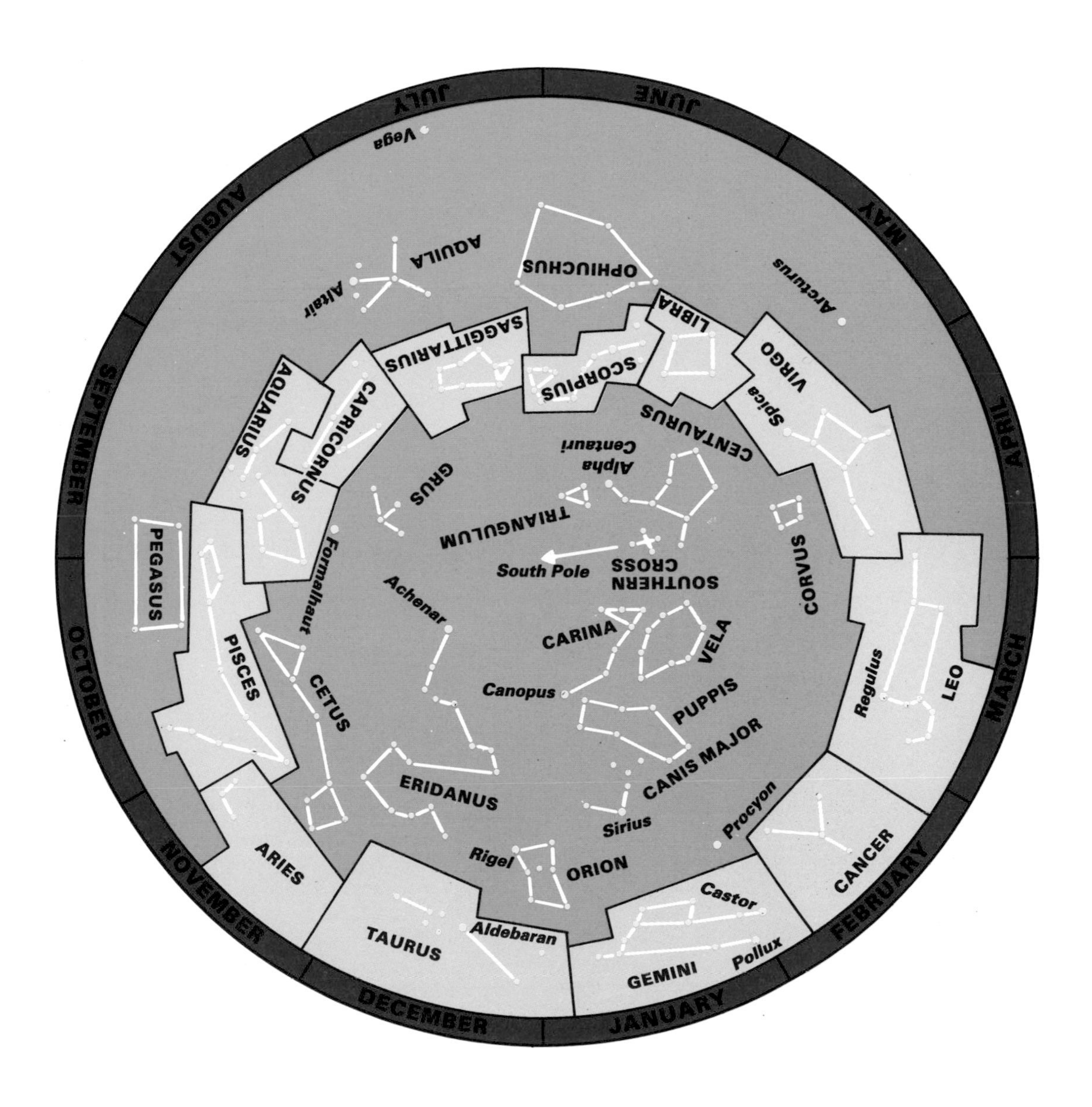

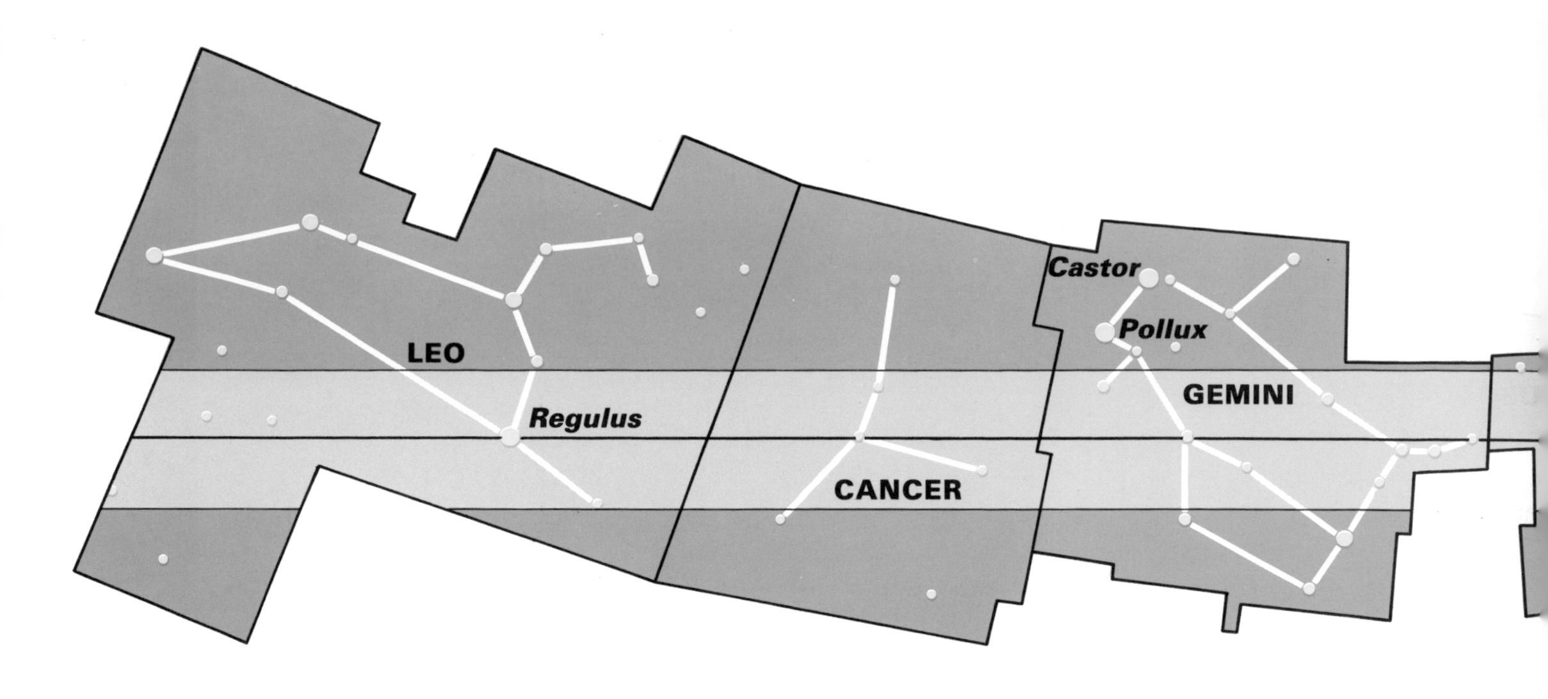

MARS

1985	October–December	Virgo
1986	January	Libra
	February–March	Scorpius/Ophiuchus
	April–September	Sagittarius
	October–November	Capricornus
	December	Aquarius
1987	January–February	Pisces
	March	Aries
	April–May	Taurus
1988	February	Scorpius/Ophiuchus
	March	Sagittarius
	April–May	Capricornus
	June	Aquarius
	July–December	Pisces
1989	January–February	Aries
	March–April	Taurus
	May–June	Gemini
1990	March	Sagittarius
	April	Capricornus
	May	Aquarius
	June–July	Pisces
	August	Aries
	September–December	Taurus

SATURN

1985	January–August	Libra
1986	February–September	Scorpius/Ophiuchus
1987	February–September	Scorpius/Ophiuchus
1988	March–October	Sagittarius
1989	March–October	Sagittarius
1990	April–November	Sagittarius

VENUS

	Morning	*Evening*
1985	May-December	
1986	November–December	February-October
1987	January–July	October-December
1988	July-December	January–June
1989	January–February	May–December
1990	February–September	December

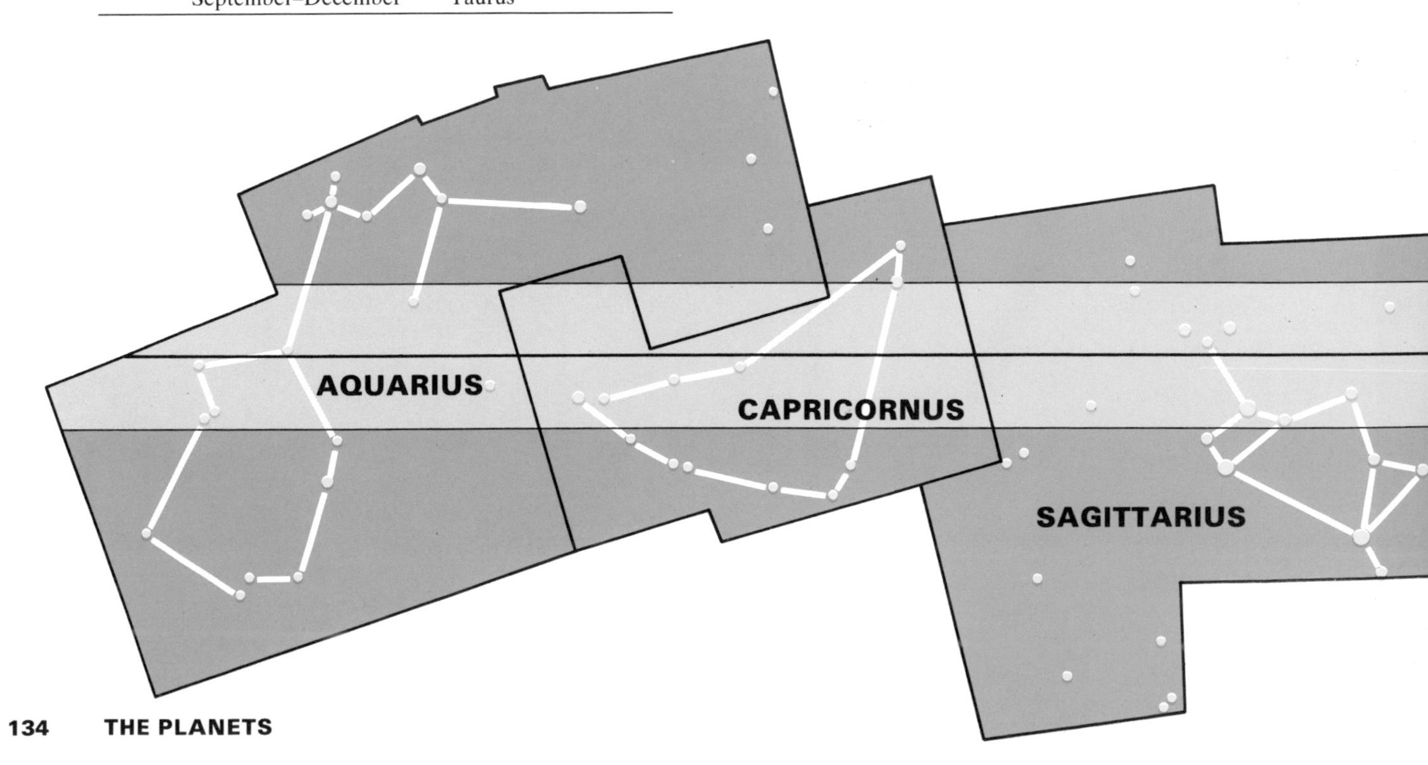

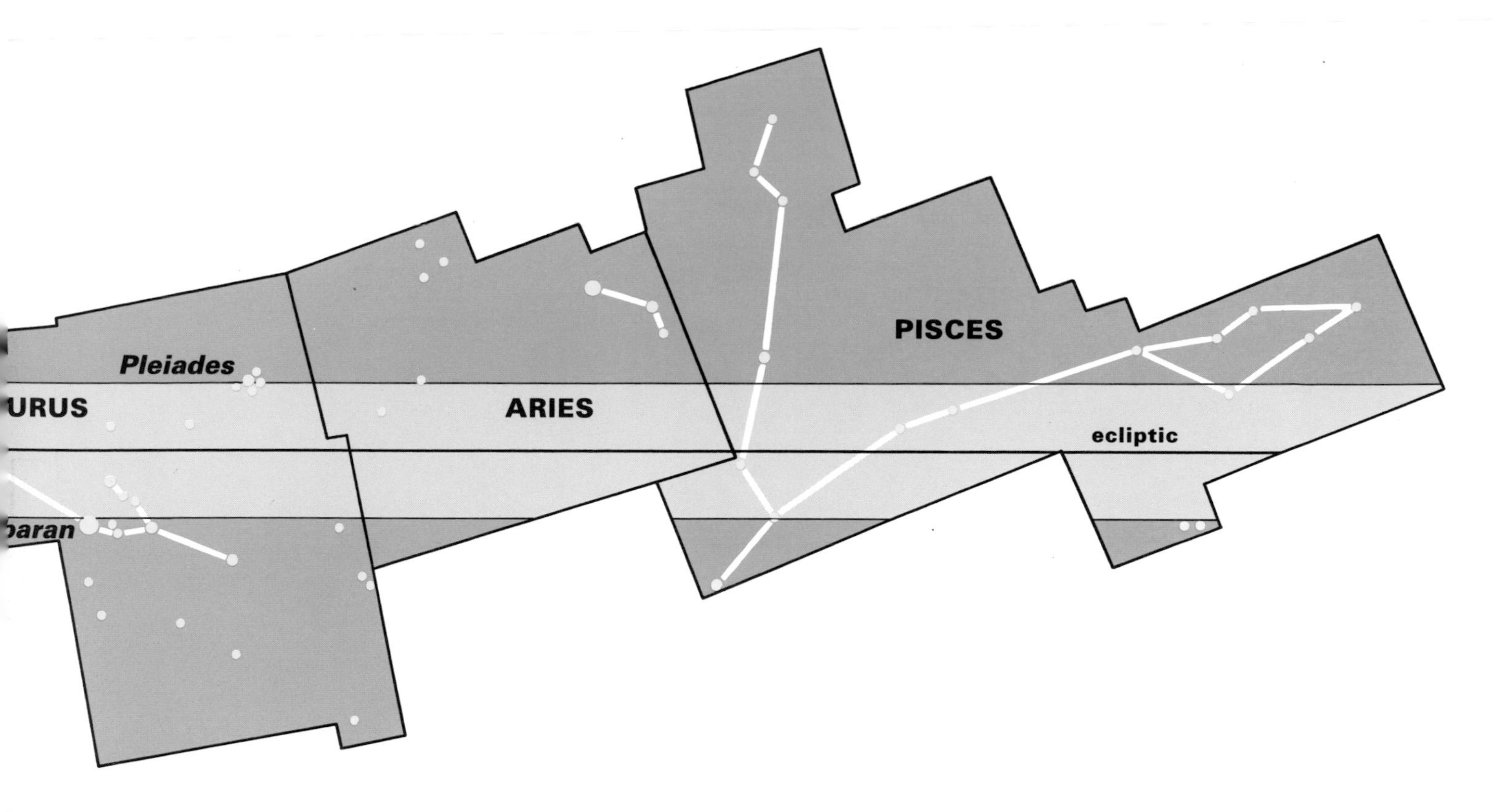

MERCURY

	Morning	*Evening*
1985	August–September	June–July
1986	August November–December	February–March June–July
1987	 November	January–February May–June
1988	February–March October–November	January May
1989	 October	January April–May December
1990	February May–June September	April August December

JUPITER

1985 **1986**	March–December January	Capricornus Capricornus
1986 **1987**	May–August September–December January–February	Pisces Aquarius Aquarius
1987 **1988**	June–July August–October November–December January–March	Pisces Aries Pisces Pisces
1988 **1989**	July–December January–April	Taurus Taurus
1989 **1990**	August–December January–May	Gemini Gemini

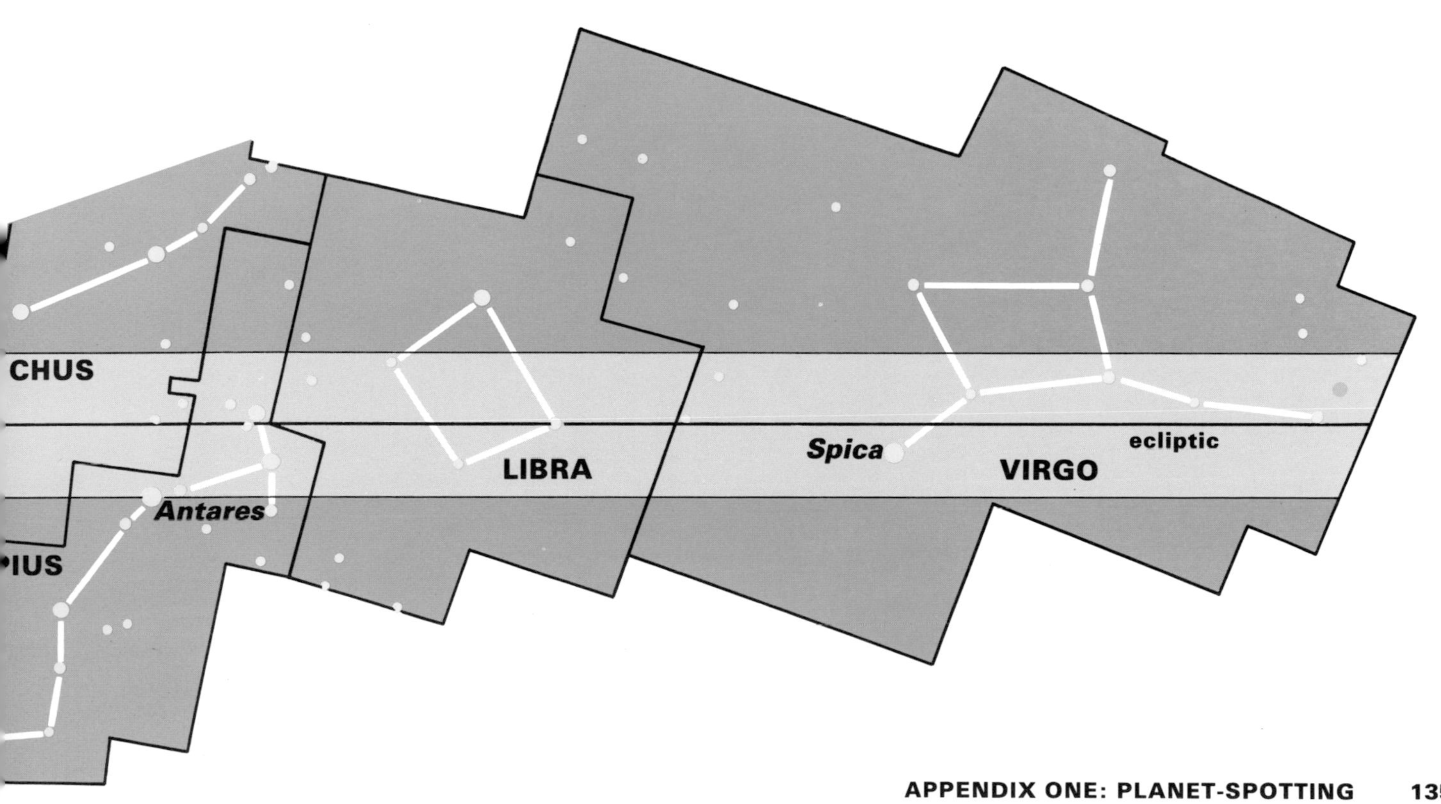

APPENDIX TWO

OBSERVING THE PLANETS

Apart from the satisfaction of recognizing a planet and following its path in the sky from week to week, there's little more that can be done with the naked eye alone. To observe a planet, rather than just see it, you need optical aid – binoculars or a telescope. Before you splash out, though, remember that both binoculars and telescopes show a much more limited region of the sky than your eye. So it helps to be familiar with the main star patterns first.

Binoculars are excellent for getting started in astronomy, whether you want to observe planets, stars or the Moon. They're described by their size and their magnification. For instance, a pair of binoculars referred to as '7×50' will have lenses 50mm across, and magnify seven times. Although this mightn't sound like much magnification, that's no bad thing. It's difficult to hold big binoculars steady, and hugely-magnifying binoculars (like the 25×40s you sometimes see advertised in the papers) also magnify every wobble. Your best bet is to get a pair which have the biggest lenses you can afford – they let in the most light and so reveal the most detail – and to support them well when you are observing, by resting your elbows on a table or a wall.

Only buy a *telescope* when you know you're hooked on stargazing. The price should be enough to deter you, anyway! Even the smallest telescope proves expensive. And more often than not, many small telescopes today are almost useless. Avoid like the plague any that claim 'fantastic 900× area magnification!!'. This simply means that it magnifies thirty times and someone has 'squared up' the figure to make it sound impressive.

As with binoculars, a telescope's magnification is not that important – but 'light grasp' is. Beware many small refracting telescopes which actually haven't got as big a lens as they claim. In these cases, the quality of lens is so poor that it has to be 'stopped down' to a much smaller area. So little light gets in that it's hard to see anything at all.

Once again, the moral is to buy a telescope with the biggest lens or mirror you can afford – and prepare to pay more than you reckoned. A refracting telescope with a lens less than 2½ inches (60mm) across, or a reflecting telescope whose mirror is smaller than 4 inches (10cm) is not really worth having. Invest in a good, firm stand if the telescope hasn't got one already. If you want to take photographs, you will need a mounting with a drive to counteract the effects of the Earth's rotation – but check a book on practical amateur astronomy for details of this. Finally, be prepared for a disappointment. The views you get will be nothing like as spectacular as the close-up space probe images in this book. You'll see a planet as a small, subtly-coloured disc, blurred from time to time by the shifting currents in our atmosphere. But with practice, you will learn to record the mass of detail which appears fleetingly when the air is still – when the 'seeing' is good. With this will come the satisfaction of knowing that you are actually looking in close-up at another world, with your own eyes.

MERCURY is such an elusive planet that it's hard to spot unless you know exactly where to look. Usually, it lies too close to the Sun to be easily seen. But on spring evenings you can spot it low in the west after sunset, and it can also be seen on autumn mornings before sunrise. However, it's best to check first with a yearbook of astronomy, or with the BAA Handbook (see below), before spending fruitless hours searching. Through *binoculars*, the view isn't much better. Even in a *telescope*, Mercury is disappointing. It does show a sequence of phases, like the Moon or Venus, but they are hard to see in a telescope less than 6 inches (15cm) in diameter.

VENUS is quite the opposite of Mercury. It can be a dazzling object in the dusk or dawn sky, and you can't confuse it with anything else. *Binoculars*, if supported well, may show a tiny crescent when Venus is at its closest. Through a *telescope*, Venus can be overpoweringly bright. It's best to observe the planet against a twilight sky when Venus appears dimmer. A telescope will reveal the whole cycle of phases, and a large telescope may show hints of markings on Venus' cloud-tops.

THE MOON shows a surprising amount of detail even to the unaided eye. If you catch it against a daytime or twilight sky, before it becomes too bright, you can pick out the big impact basins where the astronauts landed ('the Man in the Moon'), and you can even see some of the more prominent craters, like Copernicus and Tycho, as bright spots. And, of course, you can follow the ever-changing phases. *Binoculars* reveal a great deal – mountains, craters and plains. And even the smallest *telescope* can give you the feeling of 'being there'. A 6 inch (15cm) diameter telescope will reveal details on the Moon as small as a couple of miles (3km) across. If you have never observed before, the Moon is by far the best place to start.

MARS can sometimes be confused with a star, but its vivid red colour gives it away. To the naked eye, it looks like a point of light; but even without a telescope, you can follow its path in the sky and watch it doing the occasional backwards loop as the Earth overtakes it. *Binoculars* reveal nothing on Mars; its disc is just too small. And you need a *telescope* at least 6 inches (15cm) across to see very much on the little planet. But a medium-sized telescope will show you Mars' polar caps and the dark markings. These shift slowly around the disc as you watch because of the planet's 24½-hour rotation. If you do become a keen Mars observer, you will be able to follow the seasonal growing and shrinking of the polar caps and the dark markings, and perhaps catch the occasional dust storm obscuring its features.

JUPITER is the second-brightest planet, and it shines with a creamy-yellow light. It's on record that some people can actually see its four 'Galilean' moons with the unaided eye. You can certainly see them through *binoculars*, provided the binoculars are supported well. But you need a *small telescope*, at least, to watch the moons moving behind and in front of the planet, and to see Jupiter itself well. A small telescope will show details in Jupiter's cloud belts which change from day to day. You can also see the planet spinning before your very eyes – its 'day' is less than ten hours long!

SATURN can be confused with a star at first glance, so check carefully with a starchart if you want to find it. *Powerful binoculars* will reveal the rings, but you will need a pair which magnify at least ten times. In a *small telescope*, Saturn is a beautiful sight – it looks quite unreal, like a tiny model planet suspended in space. You will be able to see the rings and pick out some of the brighter moons. Telescopes more than 6 inches (15cm) across will show details in the rings like Cassini's Division, but don't expect to see much on Saturn's disc; it hides its storms under a layer of haze.

URANUS *is* visible to the unaided eye – but only just – if you know exactly where to look. Neither *binoculars* nor *telescopes* show any detail at all on its pale, greenish disc. But if you want the satisfaction of being able to identify the planet, check its position on a good starchart such as the one published in the annual BAA Handbook.

NEPTUNE AND PLUTO aren't visible without optical aid. *Binoculars* will show Neptune as a faint, bluish 'star', but you'll need a *telescope* at least 8 inches (20cm) across to see Pluto at all.

METEORS, the tiny grains of dust from dying comets, can be seen streaking through the sky every night as they burn up in our atmosphere. But at certain times of year (listed below) we see more meteors than usual – up to sixty per hour. That's when Earth ploughs through clouds of comet debris in space, and we get treated to a meteor shower. Perspective makes the meteors appear to come from one spot in the sky, and the showers are named after the constellation they seem to 'radiate' from. It's best to observe meteors without any optical aid at all; that way, you get the widest possible view. Wrap up really well, lie down on a sunlounger and just look up – that's the best way to see them!

Main meteor showers

Name	*Date of Maximum*	*Average max. no. per hour*
Quadrantids	3 Jan	40
Lyrids	21 Apr	15
Eta Aquarids	4 May	20
Delta Aquarids	28 July	20
Perseids	12 Aug	60
Orionids	21 Oct	25
Taurids	3 Nov	15
Leonids	17 Nov	15
Geminids	13 Dec	50
Ursids	22 Dec	15

COMETS, the parents of meteors, are a much rarer sight in our skies. Dirty snowballs which normally live outside the domain of the planets, they can only be seen when they come close enough to the Sun to be partly vaporized. Unfortunately, the popular image of a comet with a glowing head and a tail of gases millions of miles long isn't one that most comets live up to. The majority are so small that they never even sprout a tail – and they can only be seen with powerful telescopes.

But a few comets which come round regularly (especially Halley's Comet) *do* become visible to the unaided eye. When a comet comes close to Earth, there's usually plenty of time to observe it. Unlike a meteor, a comet doesn't suddenly streak across the sky. As it's outside the Earth's atmosphere, it only moves slowly from night to night, and *binoculars* are ideal for following its progress.

HALLEY'S COMET 1985–86

The most famous of comets returns to the Sun every seventy-six years. At its closest, it shines in the sky for several months as a glowing misty cloud with a shining tail. We'll see Halley's Comet at various periods from November 1985 to May 1986, losing sight of it when it is below the horizon, or is too close to the Sun. Unfortunately, on this return, the comet won't be very bright. Look for it where skies are very dark, away from streetlights.

In the following guide, you can identify the named constellations from the charts on pages 132–3 or 134–5. (The comet's orbit is so tilted that it can stray away from the Zodiac.)

November 1985: Halley's Comet is still too faint to see with the naked eye, but you can pick it up with binoculars as a fuzzy patch in Taurus. On 16 November, the comet lies very close to the Pleiades (Seven Sisters). The comet then moves into Aries.

December: During this month, the comet becomes bright enough to be seen with the unaided eye. It passes right through Pisces into Aquarius.

January 1986: Staying in Aquarius, the comet becomes brighter. On 13 January, it's near Jupiter and the crescent Moon. Towards the end of the month, the comet disappears into the evening twilight.

February: Halley's Comet passes closest to the Sun on 9 February, and for most of the month it is invisible in the Sun's glare.

March: The comet is too far south to be seen from northern countries like the UK. Southern observers will see the comet moving through Capricornus and Sagittarius. European, Japanese and Soviet space probes photograph the comet this month.

April: The month when the comet is most brilliant – but it is still only visible from southern latitudes. Moving from the south of Scorpius through Centaurus, the comet should boast an impressive tail. 24 April will see a spectacular sky sight, as the Moon is eclipsed while the comet is high in the sky.

May: The rapidly-dimming comet has climbed in the sky, so northern observers will see it just below Leo. By the end of May, you will need binoculars to see the comet at all – and that is your last chance until 2061!

If you want to take your interest further, perhaps to make a more systematic study of the Moon and planets, the two major amateur astronomical societies in the UK will be happy to have you as a member.

For complete beginners of all ages, there is the Junior Astronomical Society. For details, send an sae to Martin Ratcliffe, 10 Swanwick Walk, Tadley, Basingstoke, Hants., RG26 6JZ.

The main society for already active observers is the British Astronomical Association, Burlington House, Piccadilly, London W1V 0NL: it publishes the very comprehensive BAA Handbook every year. The BAA co-ordinates observations from amateur astronomers the world over, and its members have made many notable discoveries.

PICTURE CREDITS

Heather Angel/Biofotos: 37
Anglo Australian Telescope Board: 70–71, 87, 100–101, 130–131
University of Arizona/JPL: 123
G. Baier & G. Weigelt: 108
The Bridgeman Art Library/Uffizi: 22–23
J. W. Christy: 107 (bottom left)
Daily Telegraph Colour Library: 33
D. di Cicco/Sky and Telescope: 113
Paul Doherty: 122 (bottom)
Eidgenössische Technische Hochschule, Zürich: 124–125
Mary Evans: 12(top), 12(bottom), 20(bottom), 43, 86(bottom), 96, 108–109, 114
V. Fifield: 23
Peter Fisher: 13
Geoscience Features: 39
B. A. Goldberg, G. W. Garneau, S. K. Lavoie/JPL: 79 (bottom)
Hencoup Enterprises: 16–17(bottom), 100, 102(bottom), 104–105, 106, 107(top), 111, 112(bottom), 116–117(bottom), 121(top)
Michael Holford: 8–9, 34(left), 120
Jonathan Horner: endpapers
The Kobal Collection/Paramount Pictures: 56
Lowell Observatory: 104, 105
LPL/University of Arizona: 98(bottom), 102(top)
A. McEwen/USGS/NASA: 80–81
Patrick Moore: 16–17(top), 58, 121(bottom)
NAIC: 25
NASA: 2, 7, 14, 15, 18, 18–19, 19, 21, 24, 18, 30, 35, 40(top), 40(bottom), 41, 42, 44, 45, 46, 48–49, 51(top), 52–53, 53, 54(bottom), 57, 60(top), 60(bottom), 61, 62–63, 64–65, 65, 66, 67, 68, 69, 70, 75, 77, 78(left), 78(right), 79(top), 82, 83, 84, 88–89(top), 88–89(bottom), 90–91(top), 90–91(top), 90–91(bottom), 92(top), 92(bottom), 93, 98(top), 99(top), 99(bottom), 101, 110–111, 116–117(top), 128–129(top), 128–129(bottom)
NASA/SPL: 29, 55, 62
Reproduced by permission of the Trustees of the National Maritime Museum: 94, 95(left), 95(right)
Novosti: 31, 51 (bottom)
NRAO: 72(top)
NRAO/I. de Pater: 72(bottom)
RAL: 112(top), 122(top)
Ann Ronan: 10, 11(left), 11(right), 20(top), 34(right), 49, 59, 74, 86(top), 96–97, 118–119, 119
© Royal Observatory 1980, Edinburgh/Photolabs: 127
Sheridan Photo-Library: 16, 74–75, 84–85, 94–95, 103
Soames Summerhays/Biofotos: 38
TASS: 26–27(top and bottom)
US Naval Observatory: 107(bottom right)
R. Wolff/JPL/Trend Western Technical Corp.: 73

APPENDIX THREE

FACTS AND FIGURES

PLANETS

PLANET	Diameter at equator miles (km) and relative to Earth	Mass relative to Earth	Density relative to water	Average temperature °C	Tilt of axis	Average distance from Sun millions of miles (km) and A.U.*	Range in distance A.U.*	Period of revolution ('year')	Period of rotation†	No. of known moons
Mercury	3031 (4878) 0·38	0·055	5·5	350 (day) −170 (night)	2°	36·0 (57·9) 0·387	0·31–0·47	88 d	59 d (176 d)	0
Venus	7520 (12,103) 0·95	0·81	5·2	465	2°	67·2 (108·2) 0·723	0·72–0·73	225 d	243 d (E to W) (120 d)	0
Earth	7962 (12,756) 1·00	1·00	5·5	15	23·5°	93·0 (149·6) 1·000	0·98–1·01	365 d	24 hr	1
Mars	4222 (6794) 0·53	0·11	3·9	−23	24°	141·6 (227·9) 1·523	1·38–1·66	687 d	24 hr 37 m	2
Jupiter	88,730 (142,800) 11·0	318	1·3	−150	3°	483·6 (778·3) 5·202	4·95–5·45	11·9 yr	9 hr 55 m	16
Saturn	74,560 (120,000) 9·41	95	0·7	−180	27°	886·7 (1427·0) 9·538	9·00–10·01	29·5 yr	10 hr 40 m	20+
Uranus	32,560 (52,400) 4·11	15	1·3	−210	82°	1783·3 (2869·6) 19·181	18·28–20·08	84 yr	16 hr (E to W)	5
Neptune	31,400 (50,540) 3·96	17	1·7	−220	29°	2794·1 (4496·7) 30·058	29·79–30·33	165 yr	18 hr	2
Pluto	1900 (3000) 0·24	0·002	0·5	−220	50°	3666 (5900) 39·44	29·58–49·3	248 yr	6 d 9 hr (E to W)	1

* A.U. = Astronomical Unit, the distance of the Earth from the Sun.
† Relative to the stars. If the length of the 'day' (sunrise to sunrise) is very different, it is given in brackets.

MOONS

Name	Diameter miles (km)	Distance from planet miles (km)	Discovered by	Date
Earth				
Moon	2160 (3476)	238,857 (384,403)	—	—
Mars				
Phobos	14 (22)	5,800 (9,300)	A. Hall	1877
Deimos	7 (12)	14,600 (23,500)	A. Hall	1877
Jupiter				
Metis	25 (40)	79,500 (128,000)	Voyager	1979
Adrastea	15 (25)	80,200 (129,000)	Voyager	1979
Amalthea	130 (210)	112,700 (181,300)	E. E. Barnard	1892
Thebe	60 (100)	137,900 (221,900)	Voyager	1979
Io	2256 (3630)	262,000 (421,600)	S. Marius, Galileo	1610
Europa	1950 (3138)	416,900 (670,900)	S. Marius, Galileo	1610
Ganymede	3270 (5262)	665,000 (1,070,000)	S. Marius, Galileo	1610
Callisto	2983 (4800)	1,168,000 (1,880,000)	S. Marius, Galileo	1610
Leda	5 (10)	6,893,000 (11,094,000)	C. Kowal	1974
Himalia	110 (180)	7,133,000 (11,480,000)	C. D. Perrine	1904
Lysithea	15 (25)	7,282,000 (11,720,000)	S. B. Nicholson	1938
Elara	50 (80)	7,293,000 (11,737,000)	C. D. Perrine	1905
Ananke	15 (25)	13,100,000 (21,200,000)	S. B. Nicholson	1951
Carme	20 (30)	14,000,000 (22,600,000)	S. B. Nicholson	1938
Pasiphaë	25 (40)	14,600,000 (23,500,000)	P. Melotte	1908
Sinope	20 (30)	14,700,000 (23,700,000)	S. B. Nicholson	1914
Saturn				
Atlas	20 (30)	85,600 (137,700)	Voyager	1980
'Inner shepherd'	60 (100)	86,600 (139,350)	Voyager	1980
'Outer shepherd'	50 (80)	88,000 (141,700)	Voyager	1980
Janus	120 (190)	94,080 (151,420)	Voyager (A. Dollfus/	1980
Epimetheus	75 (120)	94,120 (151,470)	S. Larson/J. Fountain)	(1966)
Mimas	240 (390)	115,300 (185,500)	W. Herschel	1789
Enceladus	310 (500)	147,900 (238,000)	W. Herschel	1789
Tethys	660 (1060)	183,100 (294,700)	G. D. Cassini	1684
Telesto	15 (25)	183,100 (294,700)	S. Larson/H. Reitsema/ B. Smith	1980
Calypso	15 (25)	183,100 (294,700)	R. Harrington/D. Pascu/ K. Seidelmann	1980
Dione	700 (1120)	234,500 (377,400)	G. D. Cassini	1684
'Dione B'	20 (30)	234,500 (377,400)	P. Laques/J. Lecacheux	1980
Rhea	950 (1530)	327,500 (527,000)	G. D. Cassini	1672
Titan	3200 (5150)	759,200 (1,221,900)	C. Huygens	1655
Hyperion	150 (250)	920,300 (1,481,100)	W. Bond/W. Lassell	1848
Iapetus	910 (1460)	2,212,900 (3,561,300)	G. D. Cassini	1671
Phoebe	140 (220)	8,050,000 (12,954,000)	W. H. Pickering	1898
(The Voyagers have photographed several more moons, but their orbits are not known.)				
Uranus				
Miranda	300 (500)	81,100 (130,500)	G. P. Kuiper	1948
Ariel	800 (1300)	119,200 (191,800)	W. Lassell	1851
Umbriel	680 (1100)	166,000 (267,200)	W. Lassell (W. Herschel)	1851 (1802)
Titania	1000 (1600)	272,200 (438,000)	W. Herschel	1787
Oberon	1000 (1600)	364,300 (586,300)	W. Herschel	1787
Neptune				
Triton	2200 (3500)	219,000 (353,000)	W. Lassell	1846
Nereid	300? (500?)	3,450,000 (5,560,000)	G. P. Kuiper	1949
Pluto				
Charon	750? (1200?)	12,000 (19,000)	J. Christy	1978

INDEX